AN INTERNET
IN YOUR HEAD

AN INTERNET
IN YOUR HEAD

A NEW PARADIGM FOR
HOW THE BRAIN WORKS

DANIEL GRAHAM

Columbia University Press   *New York*

Columbia University Press
Publishers Since 1893
New York   Chichester, West Sussex
cup.columbia.edu
Copyright © 2021 Columbia University Press
Paperback edition, 2022

Library of Congress Cataloging-in-Publication Data
Names: Graham, Daniel, author.
Title: An internet in your head : a new paradigm for how the brain
works / Daniel Graham.
Description: New York : Columbia University Press, 2021. |
Includes bibliographical references and index.
Identifiers: LCCN 2020040237 (print) | LCCN 2020040238 (ebook) |
ISBN 9780231196048 (hardback) | ISBN 9780231196055 (pbk.) |
ISBN 9780231551618 (ebook)
Subjects: LCSH: Computational neuroscience—Research. |
Neurobiology—Mathematical models. | Communication. |
Neural networks (Neurobiology)
Classification: LCC QP357.5 .G73 2021 (print) |
LCC QP357.5 (ebook) | DDC 612.8/233—dc23
LC record available at https://lccn.loc.gov/2020040237
LC ebook record available at https://lccn.loc.gov/2020040238

Cover design: Julia Kushnirsky
Cover image: Alamy
The epigraph to chapter 7 is from Santiago Ramón y Cajal, *Advice for a Young Investigator*, translated by Neely and Larry Swanson (2004), published by The MIT Press.

CONTENTS

Preface vii

1 The Internet-Brain and the Computer-Brain 1

2 Metaphors for the Brain 26

3 What We Don't Know About Brains 65

4 From Connectomics to Dynomics 101

5 How the Internet Works 119

6 The Internet Metaphor: First Steps to
a New Theory of the Brain 150

7 Critique of the Internet Metaphor 206

8 The Internet Metaphor in Action: Emerging Models
and New Technologies 240

9 The Internet Metaphor, AI, and Us 259

Afterword 271
Acknowledgments 275
Notes 277
Bibliography 315
Index 337

PREFACE

WE think of our brains as computers. Whether we notice it or not, we invoke the metaphor of the brain as a computer anytime we talk about retrieving a memory, running on autopilot, being hardwired for something, or rebooting our minds. Neuroscientists are no less trapped in the computer metaphor. For almost as long as neuroscience has been a recognized field, the default approach has been to imagine the brain as a computing device.

Of course, most neuroscientists don't think the brain is literally a digital computer. But textbooks across the brain sciences routinely describe neurobiological processes of thinking and behavior as directly analogous to those of a computer, with programs, memory circuits, image processing, output devices, and the like. Even consciousness is described as the internal computational modeling of the external world. And although comparisons of the brain to a computing device are usually somewhat qualified, they are nearly ubiquitous. The metaphor is hard to escape or even notice because it is so ingrained in the way we think about the brain.

This situation exists in part because neuroscientists use the computer metaphor when describing the brain for the general public. Neuroscientist Dean Buonomano, in his 2011 book *Brain Bugs*, calls brain injuries and disorders a "system crash," and he writes of "disk space" and "upgrades" for our memory systems.[1] Cognitive scientist Donald Hoffman analogizes our visual perception of the world with a computer desktop interface: "my icon of an apple guides my choice of whether to eat, as well as the grasping and biting actions by which I eat."[2] Others, like brain scientist Gary Marcus, are uncompromising: "Face it," Marcus wrote in the *New York Times*, "your brain is a computer."[3]

Neuroscientists typically see the job of a given part of the brain—single neurons, neural circuits, or brain regions—as computing something. At each level, electrical or chemical signals are passed among components and the components operate on the signals by computing something. Computing in this sense means taking in a signal, making the signal bigger or smaller, faster or slower, and then passing the signal along for further mathematical adjustment. What matters is the computational relationship between the magnitude of the signal coming in and the magnitude of the signal going out.

A neuron's job is often to compute a response when provided with some stimulus: a pattern of light, a sound, a social situation. With lots of neurons performing specialized computations, properties of our environment can be sensed, analyzed, stored, and linked to behavior. Working neuroscientists mostly agree that, although brains and computers differ in innumerable ways, they share a common set of "hacks." In other words, brains and computers exploit many of the same fundamental design principles.

There is no doubt that the computer metaphor has been helpful and that the brain does perform computations. But neuroscience based on the computer metaphor is incomplete because it does not consider the principles of network communication. Neuroscientists are starting to realize that, in addition to performing computations, the brain also must communicate within itself. The key point is that, although communication involves computation, communication systems rely on different fundamental design principles than those of computing systems.

Although it has been little studied, brain-wide communication is attracting greater interest. We increasingly understand the physical structure of the brain as a highly interconnected network. The connectomics movement aims to map this network, as well as its dynamic activity. Through increasingly massive studies of the structure of neuronal networks, a new picture of brain function in complex animals is emerging. We are beginning to understand that one of the connectome's main jobs is to support brain intercommunication.

At the moment, however, there is no guiding principle for how these interconnected networks carry messages to and from a given part of the brain. We don't know the rules about how traffic on brain networks is directed or how the rules relate to our capabilities of thinking and behavior. We don't even know how to investigate this. What's missing, at least in part, is an appropriate metaphor to help us think about how the brain communicates within itself. I propose that the internet is that metaphor. The computer metaphor and the internet metaphor can coexist and inform one another. For one thing, the internet is obviously made up of computers. But it has emergent properties and rules that differ from those that govern single computers.

The coexistence of computation and communication metaphors—and the change in perspective needed to understand communication strategies—can be understood as being analogous to a time traveler from the past encountering today's internet. Imagine a 1950s-era electrical engineer transported to the present day. The engineer doesn't know what the internet is, but given a standard Wi-Fi router, she is curious enough to open it up and record electrical currents from its circuit board. By carefully measuring voltage changes over time at various locations on the circuit board, the engineer could probably learn to identify different kinds of components, such as diodes and transistors. In doing so, she could deduce the computations each one performs. But the stream of voltage variations entering or leaving the router would be very difficult to interpret. Measuring only the sheer number of signals would reveal little.[4]

In brains, we have something similar. We can measure the activity of individual cells in the brain and deduce the rules that govern their electrical changes. In a much more limited way, we can measure large-scale brain activity. But we can't observe how messages are transmitted across several synapses in the brain or the branching, dynamic paths these messages may take.

In short, we don't know the brain's strategy for passing messages across the whole brain. Indeed, supposing the existence of "messages" is somewhat heretical. But returning to our time-traveling engineer, if she knew the general rules for message passing on a computer network, she might be able to identify the role played by a given diode or transistor in the router. The same should be true for brains: if we could work out the basic principles of message passing, we could understand the role of individual neural computations.

For decades, neuroscientists have been measuring diodes and transistors and ignoring the larger system of message passing. We should think more about the brain as a unified communication system in science—and in society. Going further, we can investigate the brain in reference to the general principles that make the internet the universe's most powerful, flexible, and robust communication system. This change in viewpoint can also help us all understand and utilize our own brains more effectively.

We know that brains must intercommunicate at all levels, from the biochemistry of synapses to whole-brain oscillations in electrical activity. Most importantly, it must be possible to send messages selectively in the brain without changing the structure of the cellular network of neurons. All kinds of tasks involve sending messages to one place sometimes and to another place at other times. This seems obvious when stated directly, but it is rarely acknowledged.

It's like what happens at an ice cream shop when we decide between chocolate and vanilla. It must be possible for a decision-making neuron in our brain to direct a signal to the neural output for saying "chocolate" or, alternatively, to the neural output for saying "vanilla." We might even say "chocolate—no, wait! Vanilla!" because we remember that the vanilla at the shop is especially tasty, and thereby draw upon memories stored elsewhere on the network to change the route of the message in real time. The trace of communication across the network can change almost instantaneously. But our brain accomplishes this without altering neuronal network connectivity.

Neuroscientists have extensively studied the decision-making computations occurring in neurons.[5] These neurons appear to "decide" to fire or not fire by accumulating evidence from input signals over time. But it is not known how the computed decision is routed to the selected output neurons. This question has not really even been asked.

Other parts of the body also intercommunicate, and it's worth considering whether the solutions adopted in other biological systems are useful comparisons. The immune system, for example, is predicated on the ability to pass information about the presence of pathogens to the appropriate internal security forces. Great armies of antibodies patrol every milliliter of blood, applying tiny labels to anything suspicious. As tagged microbes circulate through the body, the tags are eventually noticed and the offender pulled aside and killed. The message, as it were, has been received.

If antibodies are the immune system's messages, passed by physical movement in miles of blood vessels, the brain's messages are something altogether different. In the brain, messages consist of electrical signals and their chemical intermediaries. Messages travel over a highly interconnected—but fixed—network of "wires." No individual component of the brain moves very far, at least in the short term. It is this kind of networked message passing that defines neural communication. Just like the immune system, the brain must have effective global rules and strategies for communication. But these rules are specialized for a system made of neurons and linked to functions of thinking and behavior.

In recent years, a small but growing community of researchers has investigated the message-passing rules operating on brain networks. A few researchers have proposed internet-like solutions to the challenge of passing signals in the brain in a flexible way, though the theories have only occasionally been described as *routing* theories. *Routing* here refers to the directing of signals from one part of the network to another part according to a set of global rules. We can start to see things from a new perspective—and see how the internet metaphor can aid us—by recasting neural computation as neural routing.

• • •

The brain is not literally the same as the internet; nor is it literally the same as a digital computer. But metaphors can still be useful, and even essential. The computer metaphor has served us well in neuroscience, and there is no need to discard it. For better or for worse, it has also shaped artificial intelligence (AI), as well as popular conceptions of the brain, which we will explore later in the book. But the computer metaphor lacks the toolkit to deal with flexible communication. I will describe how the internet metaphor may begin to provide one.

In chapter 1, I delve deeper into the currently dominant computer metaphor and lay out an overview of the internet metaphor. As we will see, thinking of the brain as an internet-like system has many antecedents, but they have yet to be brought together under a single banner.

Why do we need a metaphor at all? Chapter 2 argues that metaphors have been the foundation for progress in understanding brains. But being almost inescapable, they can also lead us in the wrong direction. I conclude that good metaphors are on balance beneficial for neuroscience, and indeed for science more generally.

Chapter 3 gives a tour of the current state of research in understanding whole-brain function. Although procedural limitations on studying living brains are a major impediment to progress, our current theoretical toolkit may also be inhibiting progress. I suggest a way forward, with inspiration from holistic approaches that have already found success in some areas of neuroscience.

In chapter 4, we will examine the infrastructure for a new understanding of brains, which is being furnished by our increasing understanding of the connectome. Chapter 5 provides a primer on communication systems and the architecture of the internet, paying special attention to the features that are most relevant to the brain. A detailed presentation of the internet metaphor and its utility in neuroscience is given in chapter 6. I argue that the metaphor provides a starting point for a new theoretical understanding of the brain. The metaphor suggests novel conceptions of neural systems and offers specific mechanisms that could be instantiated in the brain.

In chapter 7, I give a critique of the internet metaphor and provide specific predictions about measurable brain processes following the internet metaphor framework. Chapter 8 looks over the horizon to new theoretical and experimental advances that could help build a fuller theory of brain function that makes use of the internet metaphor.

Finally, in chapter 9, our focus expands as we examine the wider implications of the internet metaphor that impact our conceptions of AI, consciousness, and ourselves. We will turn the internet metaphor on its head and ask a provocative question: if the brain is like the internet, is the internet in turn like the brain? In particular, could the internet be conscious? The book concludes by considering how the internet metaphor can help us use our brains more effectively in our daily lives.

AN INTERNET
IN YOUR HEAD

1

THE INTERNET-BRAIN AND
THE COMPUTER-BRAIN

HE idea of an internet-like brain may seem far-fetched or redundant in our internet-obsessed age. But going back decades, before the internet era, a few mainstream neuroscientists have used internet-based explanations for a variety of major phenomena in the brain. The first description of the brain as a flexible, internet-like message-passing system came from cognitive scientist Tomaso Poggio. "The routing of information through different channels . . . is an important problem that the nervous system must solve over and over again," he wrote in a 1984 technical report titled "Routing Thoughts."[1] Though this work is obscure today, Poggio described several major problems faced by the brain that seem to require routing explanations.[2]

Poggio focused on the visual brain. He addressed a perennial puzzle, the problem of how the brain generates visual attention. Many things draw our attention from where it is currently—new objects could appear or move suddenly, or faces could enter our visual space. We also need to be able to select one object among many in a scene. Often we shift attention by moving our eyes, but we can also shift our attention without moving our eyes at all. It is normal to change which part of our visual space is getting extra scrutiny without moving anything in our heads—and of course without changing how the eyes and brain are wired up.

How does the computer metaphor deal with visual attention? Under the computer metaphor, neuron A's job is to take in a signal—say, a dot of light shining on the retina—transform the signal, and pass that transformed signal on to all neurons that are listening to neuron A. In this view, visual attention consists entirely of increased neural activity—a computation. When we pay attention to a particular region of space, a corresponding set of neurons will become more active. In other words, the goal is to find out what computations individual neurons are performing to amplify the selected signal: what algorithm do they use? But it is not clear what attention-related increases in neural activity actually accomplish. Most importantly, these models don't explain how attention can shift dynamically and alter the flow of information.[3]

Theories of attention guided by the computer metaphor are reasonable because neurons in many brain systems do transform signals in fairly predictable ways. In some cases, we know the purpose of these transformations. But computational approaches have largely been driven by procedural necessity. In an awake animal, it is most feasible to record from just a few neurons at a time—and we are very unlikely to know how those neurons we record from are wired up, or how signals pass among many neurons.

But an overlooked part of the problem here is theoretical: classical models of neurons don't countenance the selective routing of signals. In particular, they do not permit the passing of messages to some neighbors sometimes and to other neighbors at other times. This kind of thing must occur in brains, such as when we decide between chocolate or vanilla, or attend to one part of visual space instead of another. One way this kind of flexibility could be implemented is by having local groups of neurons serve as sender, router, and receiver of signals, as on the internet. A new theoretical toolkit based on the internet metaphor allows us to imagine router-like neurons in the brain that permit the sending, transmitting, and receiving of signals across the brain. Though they have received relatively little fanfare, in recent years neuroscientists have discovered numerous potential mechanisms of this kind, which are discussed in chapter 6.

As Poggio first realized, paying attention is a job that requires flexible, selective routing. When we pay attention to something, we are choosing some part of visual space and directing its signals to parts of the brain responsible for conscious awareness and heightened visual scrutiny. At other times, different parts of visual space are sent for this extra analysis. How this is accomplished in brains is unknown, but one logical explanation is that individual neurons or groups of neurons serve as routers that direct the flow of information. It may also be necessary to increase the strength of a signal that needs attention. But the signal should be directed on the neuronal network in ways that deliver relevant information—color, shape, and movement in the area that is the focus of attention—to the correct recipients: eye movement centers in the brain, motor coordination networks, memory systems, and so on.

Another long-standing problem in neuroscience is how the visual system achieves object recognition. The ability to recognize a blob of color as a particular object is shared by practically every sighted human, as well as nearly all mammals (and indeed most vertebrates and many invertebrates). Human vision in particular seems to be built around object recognition, and all sighted humans share roughly equivalent abilities for recognition of generic objects. In other words, object recognition is something shared by a huge diversity of brains, but its basic rules don't critically depend on the particular structure of a given brain. Thus the brain's solution must be guided by some common rules of network architecture and dynamics.

Object recognition is a thorny problem. The classical theory of object recognition in our brain is largely based on the work of the vision scientist David Marr. Marr was an ardent proponent of the computer metaphor: he believed that the visual system's neural machinery was a kind of hardware that supported the software of perception. Marr didn't think the brain was literally a computer. His point was that the brain performs computations. In his landmark 1982 treatise *Vision* (published two years after his untimely death at age thirty-five), he wrote:

> Think, for example, of the international network of airline reservation computers, which performs the task of assigning flights for millions of passengers all over the world. To understand this system it is not enough to know how a modern computer works. One also has to understand a little about what aircraft are and what they do; about geography, time zones, fares, exchange rates, and connections; and something about politics, diets, and the various other aspects of human nature that happen to be relevant to this particular task. Thus the critical point is that understanding computers is different from understanding computations.[4]

Although Marr had a nuanced view of computation, his conception of object recognition left no room for communication goals or network architecture. The goal of his work was to deduce the computational purpose of each cell.

Marr's views may be in part a function of the great achievements of neurobiology in Marr's time. David Hubel and Torsten Wiesel won the Nobel Prize for pioneering research in this area. Working with anesthetized cats in the 1950s and 1960s, Hubel and Wiesel discovered that when the cat's eyes were pointed toward glowing rectangles of particular orientations, dimensions, and positions, an individual visual neuron would produce reliable electrical responses. Each cell Hubel and Wiesel measured would respond most to a rectangle of a particular shape and position. Across a larger chunk of the visual system, cells existed that were sensitive to essentially all orientations, dimensions, and positions of the rectangles in the cat's field of view. Some neurons responded to the movement of the rectangles as well, and to other visual patterns.

Hubel and Wiesel concluded that individual cells function like pattern detectors that wait around for a particular pattern of light—maybe a skinny rectangle tilted 45 degrees to the right—to cross the small patch of visual space the cell is watching. From this finding, many scientists, including Marr, concluded that the visual system's main job is to analyze each part of visual space with a vast and diverse set of these visual detectors composed of neurons.

Marr reasoned that object recognition and other functions could be extracted from the activity of these detectors using neural algorithms and computational operations. Given a predictable pattern from the world—like the pattern of light reflected from a favorite coffee mug—these systems can recognize the object from computations performed on the related pattern of neurons firing across thousands or millions of feature detectors. There are decades of evidence that this kind of system is indeed a major component of how the visual system achieves object recognition.

But this classical approach is incomplete. Because our visual world is variable, the visual system needs to ignore a great deal of variation in the visual appearance of the mug in order to reliably detect it. For example, we need to recognize the mug whether it is to the right or left of us and whether it is close or far away. In each case, the image our eyes receive may be quite different, but the required visual system output—"my mug"—needs to be the same.

Classically, there are two ways to solve this problem, which is called the problem of invariance:[5]

(1)   We could tile all of our visual space with detectors so that a complete set of mug detectors is ready across the retinal image. Whenever a pattern of light on the retina matches enough detectors in a given part of visual space, the local mug detectors activate, and we recognize it as a mug.

(2)   The other option is to find vaguely mug-looking things anywhere in our visual space. The system would then neurally shift the position of this proto-mug to a central part of visual space, and then analyze it with a common set of mug detectors.

Because option 1 would use a lot of resources in building dedicated mug detectors that might be used rarely, it is considered unlikely. Option 2, on the other hand, is favored by many scientists.[6] But a routing solution to the invariance problem requires a nonclassical mechanism: the active control of neural signals to bring the mug-like blob's visual data into register with the single set of detectors, regardless of where the mug-like blob appears on the retina.

The routing solution is by no means universally accepted, and few of those who support it realize that it was first proposed by Poggio (who used it to explain a related phenomenon). But notice that we are recasting a computation problem as a communication problem: the challenge is not to compute "mug-iness" everywhere, but rather to find vaguely muggy things on the retina and communicate their associated signals to the appropriate place at the appropriate time. Viewed from the perspective of the internet metaphor, which sees the brain as a communication system, these are precisely the kinds of models that we should be thinking about. Moreover, we can look to the internet for a vocabulary to articulate the challenges of routing. We can also look to it for basic strategies for achieving efficient routing.

Object recognition is at least partly a communication problem—not least because we accomplish it by using signals from hearing, touch, and smell in combination with vision. Attention is also a communication problem, as Poggio first saw. I will argue that learning and other core brain functions are communication problems as well. The challenges of carrying out these functions can potentially be approached using internet-like principles.

While some might see the internet metaphor as a radical departure from the computer metaphor, a look back at Poggio's ideas suggests that we might better view the internet metaphor as a strain of theoretical understanding that has long existed and is now ready for fuller expression and acceptance. Indeed, Poggio even floated the idea that the routing of information in the brain is underlain by data packets, which are the standardized chunks of information on the internet. All communications and files are divided into packets, which contain information about where they came from and where they are going. Despite the fact that he was writing in 1984, during the infancy of the internet, Poggio saw that its conception and components were deeply innovative, and he drew inspiration from it. The internet soon became a major force that would change the world—and maybe also our understanding of the brain.

• • •

The internet metaphor considers the interconnectedness of the brain to be a fundamental feature. Given its connectivity, each part of the brain can potentially send messages to practically every other part. This requires flexible routing. To quote my colleague Barbara Finlay, an eminent researcher in comparative neurology, what this amounts to is "integrating the whole damn brain."[7]

One of the most important architectural features of the internet is that there are short paths across the network connecting any user with any other user. In a technical sense, the internet is like a small-world network, where any user is only a few routers away from any other. This kind of connectedness is the same as that exemplified by the "six degrees of Kevin Bacon" parlor game: that is, the idea that "six degrees of separation" can link any actor to Kevin Bacon, or anyone on earth to anyone else.[8] The communication network of the internet, in turn, is also like a small-world network.

Brains have a similar kind of interconnectivity.[9] Connectomics has now shown that any neuron in any part of the brain is only about three or four synapses from almost any other neuron. Small-world networks are achieved not by connecting every part to every other part or by having every part connected to a central switchboard. Instead, small-world networks have high local connectivity, along with a small but significant number of shortcuts. In social networks, the shortcuts are the relatively few people with friends around the world. Shortcuts on the internet and in the brain are mostly due to a few long-range wires carrying information between more distantly separated clusters. Most connections in the brain—like most friendships—are much shorter-range, spanning only a local neighborhood. A select few communicate with peers in distant clusters, sending or receiving signals that are shared within the cluster.

Why would nature design the brain to possess such short paths between any two nodes on the network except to promote communication among essentially all parts? As brains increase in size across vertebrate species, it becomes more and more costly to build long wires for shortcuts, yet shortcuts are a universal and defining feature of the mammal brain.[10] How can a brain that has any part so near to any other part prevent activity from spilling over from one area to another? The answer, I believe, is that there is a unified routing protocol in brains—one similar to internet protocol in many respects—that coordinates the transmission of signals.

The internet's routing protocol is designed to be highly flexible. Flexibility is incredibly useful to brains as well. For example, damage to the brain and nervous system is addressed through flexibility, or what is more often called plasticity. Plasticity implies a change of shape, but in fact, when the brain is damaged, brain networks redirect the flow of signal traffic mostly on existing connections rather than changing shape (in the sense of building new connections).

An experiment in monkeys illustrates the point. In the brain, a distinct chunk of neurons is responsible for analyzing touch sensations. Each of the monkey's fingers corresponds to a particular chunk that listens to signals arising in that finger. If signals from one finger are blocked, for example by using an anesthetic, the animal can no longer feel touch in that finger. But the brain chunk that previously received those touch signals is still usable. In fact, it takes over the job of analyzing signals from the adjacent finger. With the combined brain power of the adjacent finger's original allocation of neurons, plus the neurons no longer being used for the blocked finger, touch acuity of the adjacent finger will even improve beyond what it was before. Remarkably, this change in neural traffic patterns occurs within an hour of the loss of signals from the damaged finger.[11] Since an hour is too little time to build much new wiring in the brain, brain networks must have the ability to flexibly adapt to damage and to change message flow without rewiring or altering shape.

In terms of higher cognitive function, humans are also highly flexible—perhaps uniquely so. We are the "unpredictable species," as the cognitive scientist Philip Lieberman has said.[12] We are constantly faced with new challenges, both internal and external. A flexible and volitional "mental workspace" is something we humans call upon all day, every day. Learning in particular is an inherently flexible process. Higher levels of our psyche must share information among an array of systems that generate habits, norms, reasoning, and the like. We must have a similarly flexible physical brain network to support these abilities.

The functional demands imposed on brain networks are bound to be complex. But we have at our disposal a proven approach for how to build a system that efficiently manages these kinds of demands: the internet. We know how messages are passed on the internet because we designed it. And it works! Increasingly, the internet is as reliable as the electric grid, and in some places more so. Although it has flaws, the internet works because of the flexible design of its basic architecture.

What features make the internet's routing protocol so good at being flexible? It helps to think about this question from a design perspective. Say we were designing a communication system from scratch. We would need to consider factors such as how many nodes and links there are and how they are connected. But we would also need to consider the realities of how users send messages. How many chances per day or per minute do the links get to be active? For a postal service, it is usually around once per day per address, when mail is picked up or delivered. For the internet, it is every few milliseconds or less. We would also need to ask: How big or dense a message can we send at a given time? Old-fashioned telephones are great for two-way conversation—especially of detailed or emotional information—because they are very good at capturing small auditory cues like a sigh or a variation in pitch in the speaker's voice. A phone call is good for passing information-dense messages in real time. To put it more succinctly, phones are good for synchronous two-way communication.

But if we just want to let another person know that the concert we're going to on Friday is at seven o'clock, there is no need to tie up the other person's phone by calling that person; a short email or text would be better, and it can be read whenever the recipient feels like it. That is, we would want to use an asynchronous system. We would also want to use a system that is specialized for small messages that are sparsely distributed in time, a system where there will be a brief flurry of activity over a communication channel and then a lot of silence. As we will see, the brain is, for the most part, also an asynchronous communication system, and one that behaves sparsely.

Sparse, asynchronous communication protocols of this kind keep energy costs low. This matters in the brain. The collection of neurons in our heads uses a tremendous amount of energy, so the brain must economize. Energy budgets are so restricted in the brain that the popular notion that we only use 10 percent of our brains is kind of correct: we can only have around 10 percent of our neurons firing at a given time. Most of the time it is probably less. It's not always the same 10 percent of neurons that are active (but some neurons fire rarely if ever).[13] Despite this level of sparseness in neural firing, and despite accounting for only 2 percent of body mass, the brain still consumes around 20 percent of the body's metabolism (and twice this value in infancy and early childhood).[14] The sharp restriction on energy use exerts essentially the same constraints on vertebrate brains of all sizes and should therefore be a fundamental constraint on routing protocol.[15] As we will see, the internet is also highly efficient in terms of energy use, and this is in part because activity is sparse.

Another consideration in designing a communication network is how to deal with errors. All communication systems have errors: links are damaged; too many messages arrive at once; switchboards fail; and signals are lost or corrupted. All major communication systems also have means of redressing these errors as a critical design feature. For traditional telephony, the most common error is when a caller tries to reach a party who is already on another call and gets a busy signal. The busy signal has been a ubiquitous feature of telephone communication—at least until call waiting and voicemail—since the design of the communication protocol ruled out the possibility of having two conversations at the same time.[16] Busy signals inform the caller that the line of the party called is in use and that the caller should try again later. In telephony—and in face-to-face vocal communication—errors can also be corrected once synchrony is established. If I didn't hear what you said, I can ask you to repeat yourself. Errors are routine and expected, and quickly redressed using simple solutions.

In contrast, avoiding errors in the first place is key on the internet. We can't have some parts of a message about the concert arrive (say, "Friday") and others not arrive (say, "7 p.m."). The problem is that the error can't be instantaneously noted by the receiver and corrected by the sender, as in traditional telephony; the sender and receiver are not in sync. However, because the recipient doesn't know when we pushed *send*, there is a period of time for network intermediaries to fix any errors. Routers on the network therefore make sure that every part of the message shows up in our inbox before we are notified of the message's existence.

The internet has a clever way of accomplishing this. It involves automated messages between routers. The receiver of a tranche of signals relays a message back to the sender saying, essentially, "I got it." The return-receipt messages are called *acks*, short for *acknowledgments*. Acks travel on the same network and according to the same rules as the message itself, but they can potentially take a different route back to the sender. If the sender's router doesn't receive timely acks, it will resend the missing message parts. Unlike the error correction mechanisms in telephony, which above all assume synchronous communication, the internet's acknowledgment system is asynchronous. It has attractive parallels in the brain.

The directed nature of communication on the internet is also a key design element, and also one relevant to the brain. By *directed* I mean that signals can flow in only one direction on a given channel. The idea that neurons provide directed communication originated in the early twentieth century at the birth of modern neuroscience. Based on anatomical studies of neurons, Spanish scientist Santiago Ramón y Cajal concluded that information flows only in one direction through a typical neuron.[17] As elaborated by later work, the unidirectional nature of neurons has shaped brain design in fundamental ways. But we must also consider the nature of a massive network architecture composed mostly of one-way neurons. We can see an example of the kind of network architecture the brain uses by looking at an area of the brain called the thalamus.

In mammals, there are always what Barbara Finlay calls *core brain regions*, or areas whose connections span most of the brain.[18] One of these core regions is the thalamus. It is part of the network backbone in the brain-as-internet. The thalamus holds a central place in three respects. First, it has a central position in the cranium (see figure 1.1). Second, it is central in terms of functionality. The thalamus is the indispensable gateway of vision: it is the only place in the brain that receives visual signals from the eyes leading to visual awareness. The thalamus plays a similar role of gatekeeper in all other sensory modalities except smell. It is also involved in other systems besides sensation and perception. It participates in motor coordination and regulates things such as posture. In short, the thalamus is functionally central. Third, in terms of connectivity, the thalamus has connections that send signals to a vast array of brain areas. Many of these areas have connections that reach back again to the thalamus. In fact, this system of feedback utterly dominates the data flow of the thalamus.

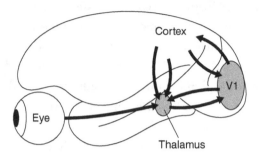

FIGURE 1.1 Information flows from the eye to the thalamus to the cortex, then back to the thalamus (and on to the cortex again). About 5 percent of the neural inputs to the main visual area of the thalamus come from the eyes; the rest come from the cortex, including primary visual cortex (area V1) and numerous other areas of the cortex as well as other parts of the brain. Looping connections are a major source of network structure in the brain's visual pathway, which could support internet-like mechanisms of flexible network communication. Image by Daniel Graham.

The flow of visual information into the thalamus is a bit like trying to watch a baseball game on a small TV while a roomful of people simultaneously yell their opinions about the game at us. All of the neurons coming from the eyes that connect to vision-related areas of the thalamus constitute only around 5 percent of the inputs to these areas. The rest of the inputs come from elsewhere in the brain. In absolute terms, the parts of the thalamus involved in vision receive inputs from roughly 2 million axons (1 million from each eye). But the same areas receive inputs from up to 40 million axons from elsewhere in the brain—they come from the cortex, the brain stem, and other places. It is hard to understate the paucity of inputs from the eye that give rise to visual awareness: everything we will ever see is delivered to the thalamus by about 0.002 percent of the neurons in our brain, and these signals are greatly outnumbered by feedback from other parts of the brain.[19]

This network structure may seem odd, but it makes more sense if we think in terms of internet-like communication. It allows signals coming from the eyes to easily reach many possible target regions in the cortex. Each little chunk of the thalamus has axons that reach distinct regions of the cortex. But feedback loops going from the cortex back to the thalamus allow messages to reach many possible targets. Signals can leave the thalamus, enter the cortex, go a synapse or two in the cortex, then return to a different part of the thalamus, and go somewhere else in the cortex.

This whole process is very fast. Loops allow messages to be passed on a round trip from the thalamus to the cortex and back in as little as 9 milliseconds (an eyeblink takes more than 100 milliseconds).[20] It is plausible that these loops could support an ack system of return messages, much like on the internet. What's important is that the right information gets to the right place, and does so reliably. Loops like those passing through the thalamus have long puzzled researchers and have generated numerous hypothetical explanations. But no one has really considered communication goals or an acknowledgment-like system.

The conventional wisdom is that the thalamus's job is to "adjust the weights" of the signals headed to the cortex, increasing some signal strengths and decreasing others-to accomplish visual attention, for example. This explanation is fully in accord with the computer metaphor, in which the job of a neuron is to compute output activity based on some mathematical function of the input. However, adjusting the weights fails to explain why loops in particular are required. Why not have local circuits in the thalamus that learn optimal weights and that operate directly on signals from the eye? Why go to the trouble to build lots of costly axons to send signals to the cortex and back? Learning optimal weights without loops to the cortex would conserve on wiring costs and save precious milliseconds of transit time. It would also be less complicated to build.

Following the logic of the internet metaphor, on the other hand, loops and small-world networks are needed because they allow short paths to many destinations, as well as the possibility of error correction using acks. This is one example of how the internet metaphor can give us new ways of understanding existing findings and suggest new ways of approaching brain networks.

Randomness plays entirely different roles in computers and on the internet. These differences can also help us understand the logic of the internet metaphor for the brain. Random variation—noise—is anathema to computers. From the most advanced quantum computer to the simplest digital calculator, noise simply cannot be tolerated. If it were tolerated, computation could produce different results with the same inputs. Noise renders computers useless. But the internet, as it turns out, requires randomness.

This was an early discovery in the internet's history. A use for randomness was found in an internet-like radio network called ALOHANET built in Hawaii in the 1970s. Injecting random variation in the message-passing process solved a key problem: how to deal with messages bumping into one another on the network. If this happens, the network will need to send both messages again. But what if they bump into one another again? After all, they are both trying to get to the same place at the same time. Engineers of ALOHANET realized that we can virtually guarantee that they won't bump into one another a second time if we introduce a small random delay in time for both signals.

The basic solution of injecting randomness is still essential to the proper function of the internet. In turn, this insight may help us understand "noise" in the brain. No one knows exactly what constitutes meaningful signals in the brain and what constitutes meaningless noise. But if we see noise as a potential feature rather than a bug, we might have a better understanding of what is actually the signal. It just takes a shift in our metaphorical frame and in our assumptions about how the complex system of the brain should operate.

In addition to highlighting the benefits of randomness, the internet metaphor helps us see the importance of interoperability in the brain. On the internet, many applications can operate over the same set of connections, and their associated signals can be intermixed as they travel across the network. This fact leads us to another general strategy on the internet: messages are, for the most part, treated equivalently. They are routed based on local traffic according to a common set of global rules. At any given moment, the data stream coming into our Wi-Fi router will be a mix of message types: a part of an email may arrive after a part of a video, followed by an ack from a previously sent email.

The system deals with this flood of messages by imposing something called a protocol *stack* on signals. The stack sorts message parts and delivers them to the correct application such as email or web. The stack is hierarchical: bare electrical signals are at the lowest level, and these are sorted into digital characters, then into data packets, then into flows of packets, and ultimately into whole messages or files. At the origin of the message, a single intention—say, writing an email—gives rise to an increasingly complex and fine-grained data representation as it descends the stack. These data are fed into the router network. Once packets arrive at the destination, the signals ascend the receiver's stack. Here, the inverse process occurs, building up dispersed signals into increasingly organized data representations.

The brain is also a hierarchical communication system whose messages are the sum of interactions at many levels. In discussing neural message passing, neuroscientist Tony Bell uses the term *emergence* to describe many-into-one behavior and the term *submergence* to describe one-into-many behavior.[21] This idea maps nicely onto the stack metaphor. From the level of receptor bio-chemistry emerges the dynamics of dendrites at the synapse, from which emerge changes in the cell's electrical properties, from which emerges a signal from the cell body to the axon to initiate a "spike." Submergence, as Bell describes it, flows in the opposite direction, with a single spike giving rise to increasingly complex and more fine-grained dynamics as the spike approaches axon terminals at the synapse. Emergence and submergence can be extended to neural assemblies and whole-brain processes. These processes are constantly interacting and conspiring to pro-duce brain activity and, ultimately, behavior. The dynamics of these systems thus resemble those of the internet protocol stack: each signal has a complex but mechanistic relationship with some larger assembly of communicated information (a "message").

As with the time-traveling engineer mentioned in the pref-ace to this book, we would have a hard time recognizing that a data packet on the internet is meaningful unless we understood the system's routing protocol. Network activity in the brain also reflects a need to organize the flow of messages that are struc-tured at many scales. With better understanding of the con-straints on brain routing protocol, we can potentially understand what a "message" is in the brain and what function a given mes-sage serves.

Furthermore, we know that neurons exchange different types of messages using common neural machinery. When we consid-ered decision-making in the brain, I noted that there are neurons whose job appears to be the weighing of evidence. Once enough evidence is gathered—by referring to past choices, for example—these cells initiate a behavioral decision based on computations on that evidence. But what is the nature of the evidence? It could be visual or auditory signals happening in real time, but it could also be memories, emotional associations, or other types of signals.

The brain must be intermodal and interoperable. Messages from different subsystems need to be able to interact with one another according to common rules. And fast. The need for interoperability is also evidenced by the very short network distances between brain parts that perform different functions. A unified protocol stack is thus a plausible conceptual model—and, perhaps, a mechanistic model—for brain-wide communication.

Once we start to see that a primary job of neurons and groups of neurons is to communicate with different targets at different times—and with neurons that are not their immediate neighbors—important features begin to emerge, sometimes unexpectedly. One crucial aspect of routing has to do with pathfinding, or figuring out how to best use the links in a communication network. On the internet, routers learn short routes to destinations from lists that are promulgated autonomously from other routers. These lists, called routing tables, are regularly updated based on probes of the network. Although many routes would potentially work well on the internet's small-world network, packets intended for the same destination are generally routed on the same path. However, if a path develops a faulty node or becomes very busy, the sending router can easily find a different path that will be just about as good as the original path. This redundancy of paths was in fact a fundamental motivating factor in the development of distributed computer networks, whose conceived purpose was to provide robust nuclear missile command if central command nodes were destroyed.

When the network finds a good path, it has learned something: it has learned about the present state of the network and of the world. Learning in the brain can be thought of in similar terms.

Classical models of learning at the neural level invoke Hebbian learning, or the idea that neurons that are active together build stronger connections.[22] This mechanism is proposed as a way to encode a pattern of response into our neurons while we learn a task. It is fully based on the computer metaphor. The notion that learning is instantiated by increasing the synaptic strength between neurons from repeated firing, which forms the basis of what we call today Hebbian learning, was proposed by Jerzy Konorski in 1948. Konorski proposed the idea based on extrapolations from known neuron properties, rather than from direct observations. It was later elaborated by Donald Hebb, whose name is associated with it.[23] The idea is that "neurons that fire together wire together." This proposed mechanism of learning has shaped neuroscience research for decades.

There is little doubt that neurons that are active at the same time tend to build stronger connections, especially during neurodevelopment. But on its own, Hebbian learning cannot explain most features of learned behavior and knowledge in the brain. There is increasing skepticism that Hebbian learning underlies memory in adult animals.[24]

Instead, a sensory experience or semantic knowledge could be learned by finding effective paths among many relevant neural units—say, between frontal decision-making areas, association areas, and sensory areas—and increasing traffic along these paths as needed. When something new is learned, a new stable path is established. Learning of this kind may require that some neurons have the ability to direct messages on a path of more than one hop, which currently is not a widely accepted idea. But brain architecture and the brain's functional demands strongly imply the possibility of multi-hop communication, which must also be flexible.

As I have emphasized, our cognitive faculties are, perhaps above all else, flexible. We can essentially choose what brain parts we want to be in communication with each other and which connections should be given more traffic at any given time.[25] We can also switch tasks rather easily. Task switching is a defining behavior of complex brains and one humans in particular excel at. But task switching is something for which computers—and especially artificial intelligence (AI)—are ill-suited. The internet has efficient solutions to this kind of problem, including the ability to reroute network traffic to new paths and the ability to carry many types of messages with varied content.

Even when the brain isn't actively cogitating or learning, it may still need to have an up-to-date knowledge of paths between essentially any node and any other node. The internet does this by promulgating routing tables among nodes. This process could correspond to the default mode network, which appears to maintain and support "background" brain function. Viewed through the lens of the internet metaphor, this activity could ensure that any brain part is potentially reachable. As we will see in chapter 6, synchrony in brain oscillations could be the mechanism for engaging paths when they are needed. A sending population of neurons that is synchronized with downstream neurons can influence those downstream neurons. Signals that arrive at the target at the wrong time are ignored.[26]

• • •

A final feature of brains that the computer metaphor ignores is that brains grow and evolve. In daily life, we would never talk about growing our laptop or phone to make it do something new—we would instead upgrade, which implies replacing one thing with another. In contrast, we often talk of growing our networks, be they meatspace social networks or computer-mediated networks, to make them more powerful. Brain networks also scale up during development and during Darwinian evolution. In the process, they acquire new functions and power.

The main challenge of scaling is to add new neurons and synapses to the system in a way that doesn't compromise existing network communication and that adds new communication possibilities for the expanded network. In development, network growth and change are mostly directed by genetic programs. But there is much allowance for unplanned growth and change. For example, the brain responds to new external conditions by slowly changing its network architecture. This is true over the time scale of an individual, as well as over evolutionary time. Thus, having a common protocol that applies relatively uniformly to communicators on brain networks could facilitate and coordinate scaling wherever and whenever it occurs.

The internet was designed for unplanned growth: anyone can join as long as they meet the straightforward, publicly available standards for interconnection. This planned capacity for unplanned growth has allowed the internet to grow faster than any other communication system in human history. And it continues to grow, accelerated in part by the coronavirus, which has shifted many meatspace activities to the online world. Today there are more internet hosts than there are neurons in the cortex of many primate species (several billion). As the core internet and the Internet of Things expand, the network of nodes may approach the size of our own cortex (over 20 billion neurons) and perhaps one day the size of our entire brain (86 billion neurons). The ways the internet has supported network scaling over long and short time scales thus may hold clues to how the brain solves similar challenges in development and evolution.

Scaling of brain networks may even offer us a way to empirically evaluate the routing strategies of brains. Different routing strategies—for example, traditional telephony versus the internet—scale in different ways. Thus, if we treat the brain as a black box routing system and probe it in the right way, we might be able to confirm that the brain uses an internet-like routing protocol, as opposed to another class of protocol. We will examine this proposal in detail in chapter 6.

$$\bullet \quad \bullet \quad \bullet$$

The brain differs from the internet in countless ways: the brain is not strictly digital; it has no common clock and operates on far slower time scales; its transmission costs are high; and so forth. Today, even "the internet" of textbooks and protocol specifications is in some ways different from the actual networked computer system of the same name. Superfast computers, in the form of routers, have rendered some aspects of internet routing protocol that ensure robustness almost redundant.

But the fundamental design choices of the internet remain critical, and they provide the most useful point of comparison for the brain. There is no internet without small-world network structure; there is no internet without packeted, interoperable messages; there is no internet without the possibility of flexible routing; there is no internet without acks; there is no internet without the assumption of asynchrony in message passing; and there is no internet without the sharing of information about how to get messages to their destination. I believe the brain uses strategies like those of the internet to solve similar problems, although it surely balances them against many others, including computational constraints.

Ultimately, the brain may use a routing protocol that differs in significant ways from the internet. There may even be several very different routing protocols in use in different brains, though I hypothesize that there is a single, unified, but flexible protocol in all mammals. The point is that the brain must have some kind of routing protocol, and neuroscience should try to figure out what it is. The internet metaphor can help us in this endeavor.

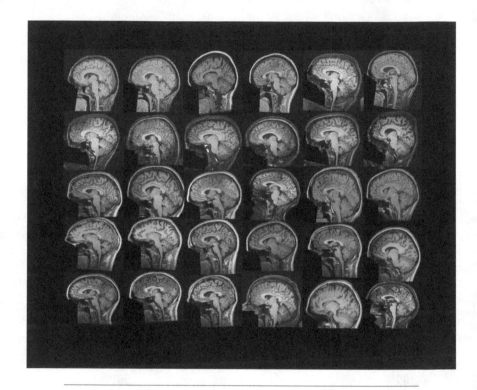

FIGURE 1.2 Shown are structural MRI images of thirty brains of adult humans in sagittal section. Note the substantial variation in shape and structure from brain to brain. Despite these differences, all are healthy individuals who share highly similar capacities for conscious experience, sensation and perception, emotion, memory, and a host of other functions. Image by Daniel Graham and Lauren Pomerantz using public data from openfmri.org.

The brain is approachable as a quantitative science by virtue of the fact that different brains can perform essentially the same functions (see figure 1.2). This means that there must be rules that all comparable brains follow in order to perform their many functions, regardless of the fine details of a given brain. Some of these rules span evolutionary lineages. I argue that the human brain's capabilities are due at least in part to principles of network structure and activity that support flexible brain-wide communication. I further argue that the internet metaphor offers a useful framework for investigating these shared principles, providing both general analogies and specific engineering solutions that can advance our understanding of the stuff going on in our heads.

2

METAPHORS FOR THE BRAIN

WHY do we need a new metaphor for the brain? Why do we need a metaphor at all? Fields like chemistry don't seem to require guiding metaphors. The idea of an acid or a base isn't a metaphor: it's a physical property of a solution that can be measured precisely in terms of pH; the terms *acid* and *base* are just shorthand. Why does biology in general and neuroscience in particular seem to need metaphors, often technological metaphors? Is this a good thing?

The short answer is that, in the physical sciences, theorists use models that are very precise metaphors. The theory behind physics, for example, is at its most basic a toy universe that we construct in our heads and express in mathematics. But the toy we imagine behaves just like the world in all observable ways, and it predicts real-world events very well. No one has ever observed an electron spinning. Yet treating an electron as an imaginary toy—a tiny bar magnet that spins[1]—allows us to build a model that successfully predicts everything about the electron that we can measure, like its interactions with other particles. Scientific theory is not identical to the world it describes. If it were, it would be useless. Its job is to expose mechanisms of action and their origins.

A metaphor in science is simply a vague theory, or the beginning of a theory. It is most useful if it is deep. The target of the metaphor should have many layers that are formally articulated and can be compared in detail to the system under study. The deeper the metaphor, the closer it is to theory. The goal is to capture key aspects of the system.

Metaphors are unfamiliar in the physical sciences because the reigning metaphors are precise enough to be couched in terms of mathematics. But metaphors—and especially technological metaphors—have been critical in the history of science, and they will continue to be so as we get closer to understanding the brain.

The Enlightenment of scientific thought that started in the seventeenth century depended heavily on the innovative technology of that era. New crafts and industries provided tools for scientists. Better optics allowed the curious to see the very far away and the very small. New hydraulic systems could create a vacuum, which had many interesting properties. But the engineering of the early Industrial Revolution also gave us metaphors. Most importantly, it gave us the notion of technology itself.

The word *technology*, though borrowed from ancient Greece, was first used in its current meaning in the seventeenth century. Linking the world's workings to technology implied that the universe could be a mechanism like a gloriously complex mechanical clock, one perhaps fashioned by God. Mechanisms, unlike metaphysics, could be studied through experimentation because they worked according to fixed, observable rules. All we need to do is to imagine a possible mechanism in the first place, then look for evidence of it. Reductionist science from then to now has largely been built on this premise.[2]

In biology, technological metaphors are ubiquitous. The clock is one of the most popular. No time-keeping systems in the body—sleep/wake or menstrual cycles, for example—work in the same way as a mechanical clock, an hourglass, or a water clock. Nor do any known biological clocks perpetually count out uniform intervals. Instead they involve chemical processes with certain periods of oscillation or decay under particular chemical and thermal conditions. Intervals need not be uniform: "tick" states and "tock" states aren't necessarily the same length in biological clocks, as they are in mechanical clocks. But the metaphor is still very useful. In part, the utility of this metaphor comes from its wide relevance in different systems in biology. For example, as we age, our DNA acquires a crust of small chemicals, a process called methylation. These chemicals, from the methyl group, glom onto the DNA molecule along the rails of the DNA ladder. They do so at a predictable rate: the degree of methylation has recently been found to be strongly correlated with chronological age, and it has therefore been termed an epigenetic clock. Again, no aspect of DNA methylation resembles a mechanical clock, but invoking the metaphor makes our understanding of the larger biological system more powerful.[3]

Technological metaphors have been particularly important in neurobiology. As with the physical sciences, it was the earliest and most basic technological metaphor that has been most important for neuroscience, and for the same reason.

René Descartes was the first to explicitly build a metaphorical linkage between natural phenomena and the hi-tech engineering of his day. In the mid-seventeenth century, he introduced the idea that living systems are mechanical automata. Descartes argued that each organism ran according to fixed rules. This insight opened new theoretical approaches, especially those that discounted the role of divine or celestial forces. If an animal is a complicated mechanism that runs on its own, the divine is less salient.

Famously, Descartes could not fully dispense with divine forces in biology. His distinction between the mystical human mind and our physical body bears his name today: Cartesian dualism. Descartes could tolerate mechanical rabbits and dogs, but humans and their seemingly special form of consciousness still required the supernatural. For better or for worse, Cartesian dualism and its implications would shape brain science through to the present day. Some have argued that what we call "cognition" today is nothing other than the mystical mind of dualism.[4] But at least in terms of the "body" half of the dualism—including the brain—Descartes's basic metaphor holds up today.

Within the broad metaphor of animals as mechanical automata, Descartes imagined a more specific metaphor for how the brain works. It was centered on plumbing. Plumbing is so pedestrian today as to be invisible. It seems to have existed forever.[5] But even in 1940, almost half of all U.S. households lacked full indoor plumbing.[6] In Descartes's time, grand waterworks like those at the Palace of Versailles were a major engineering advance, one available only to the richest (see figure 2.1). The acres of water gardens at Versailles were constructed mostly in the second half of the seventeenth century for Louis XIV of France. Continuous flows of water were delivered and distributed over 35 kilometers of piping and artfully expelled in arcs reaching several meters. All the water had to be moved uphill more than a vertical half-kilometer from the River Seine; 250 pumps drew water from the Seine, powered by the river's current. The pumps supplied more water than was delivered at the time to all of Paris.[7] The Versailles waterworks seem to have been built in part to scare foreign adversaries with France's engineering prowess.

In *Treatise of Man*, Descartes writes that our volitional minds work "just as you may have seen in the grottos and fountains of our King [Louis XIV], in which the simple force imparted to the water in leaving the fountain is sufficient for the motions of different machines, even making them play musical instruments,

FIGURE 2.1 Water gardens of the Versailles palace, France. René Descartes was among the first to make a metaphorical linkage between contemporary high technology and mechanisms of brain function. He argued that the brain may employ hydrological mechanisms like those at Versailles.

Source: "Fountain in the Parc de Versailles" by Edwin.11 (CC-BY 2.0).

or speak words according to the diverse disposition of the tubes conducting the water."[8] Descartes highlighted a part of the brain we now call the pineal body as the master valve of this plumbing system. The pineal body is unpaired, meaning it is a single, central mass, rather than a pair of mirror-image structures on either side of the brain. The pineal body is located adjacent to one of the brain's large fluid-filled ventricles. Today we know that the job of the ventricles is to nourish and protect brain tissue with cerebrospinal fluid. With his keen knowledge of brain anatomy, it is easy to see how Descartes would make the connection to plumbing in the pineal body and its neighboring ventricle. But the proximity of the pineal body to the brain's plumbing is just a coincidence; the ventricles don't control thinking or behavior.

Descartes's model was incorrect, but elements of the plumbing metaphor were retained and reappropriated at the founding of modern neuroscience. Charles Sherrington, winner of the 1932 Nobel Prize in physiology or medicine for his studies of reflex responses, described neurons as "valve-like"[9] structures. Neurons indeed regulate the flow of molecules in fluids, including neurotransmitter molecules and ions. Another founder of modern neuroscience, Santiago Ramón y Cajal, invoked the idea of a valve in the same context.[10] In classical as well as contemporary models of neuron firing, the end result of an electrical excitation in a neuron is a "puff"[11] of neurotransmitters—a spritz pumped into the synapse, not unlike the expression of a Versailles fountain. Modern neuroscience also recognizes the role of ion channels as a literal valve whose hydraulics concern charged ions and other molecules. Ion channels regulate the flow of molecules into and out of a cell, with flow often restricted to particular types of molecule.

Recall the idea that metaphors allow us to see the "hacks" or tricks behind a system. The brain has many tricks. Thanks in part to Descartes, we know that one of these tricks is plumbing (though Descartes was wrong about what exactly the plumbing system does).

Another trick the brain uses is computing. I argue that internet-like communication is another. There are surely others. Each of these tricks resembles its engineered counterpart to a greater or lesser extent. We can learn much—and peer through a new theoretical lens—whenever we imagine an appropriate metaphorical linkage to one of these engineered systems. Plumbing systems and the brain differ in countless ways, but the primary trick behind plumbing is indeed one of the tricks behind brains passing information. Plumbing doesn't explain everything. That's why we need other metaphors.

•   •   •

With technological metaphor, more correspondence between the linked entities is helpful. Another technological metaphor used by Descartes compared the human eye to a camera obscura, an optical device popularized in Europe starting several decades before Descartes's time.[12] A camera obscura is a darkened room with a tiny hole in one wall that makes an image on the opposite wall of the surrounding world. A dark room—really just a pinhole camera—might not sound hi-tech, but the key was having a lens in the hole: this allows us to admit much more light into the dark room and thereby make a nice bright image, while at the same time maintaining sharp focus. In Descartes's time, the glass used for lenses had become more pure and lens curvature more precise. The result was a sharp, bright image—one that potentially could be used as the basis of a highly accurate handmade image.[13] Artists and architects—the IT wizards of their day—were avid users. The room could even be portable. What we call a camera today is just a miniaturized version of a camera obscura, plus a shutter and film.

The correspondence of a camera obscura with the eye is obvious today. We know that we see because light enters our eyes. But this was not always understood. Since ancient times, most people in most parts of the West believed in extramission, or the idea that the eye sends out beams of particles that detect the objects they hit.[14] This view held in Descartes's era as well. Escape from this false belief required a technological metaphor. Indeed, extramission theory was not based in reference to any mechanical system but rather on phenomenology—what it feels like to see. Our naïve impression is that our eyes (and those of other people) have a radiant, penetrating quality, and this notion led to the default assumption about vision for thousands of years.

The camera metaphor shows us that seeing is about receiving rather than sending radiation. But beyond better correspondence to the eye itself, the camera metaphor also invites us to think about eyes in terms of variables relevant to cameras, such as focal length, aperture, lens curvature, and so on.

The power of a metaphor can be seen by contrasting Descartes with contemporary thinkers working from a similar base of empirical knowledge. Compare Descartes's description of brain function with that of Thomas Hobbes:

> All which qualities called sensible are in the object that causeth them but so many several motions of the matter, by which it presseth our organs diversely. Neither in us that are pressed are they anything else but diverse motions (for motion produceth nothing but motion). But their appearance to us is fancy, the same waking that dreaming. And as pressing, rubbing, or striking the eye makes us fancy a light, and pressing the ear produceth a din; so do the bodies also we see, or hear, produce the same by their strong, though unobserved action.[15]

Hobbes and Descartes both rightly rejected extramission and similar classical views. But reading the two thinkers' ideas about brains today, it is clear that Descartes's description was more influential because he adopted a concrete technological metaphor. Hobbes just talks about things bumping into each other.

Hobbes shows that a basically correct but uninspired theory is not so useful. But theories—especially in neurobiology—can also be too literal-minded. In particular, we may be attracted to theories of the brain that are built on known biological properties, but in relatively unimaginative ways. Skipping ahead two hundred years, consider Charles Darwin. His great breakthrough in evolutionary theory, as science writer David Quammen suggests, was the adoption of a tree metaphor to organize the jumble of observations he made on the voyage of the *Beagle* in South America. The metaphor coalesced in Darwin's famous notebook sketch of what became known as the tree of life, inscribed with the pregnant phrase "I think" (figure 2.2).[16]

Yet Darwin was less imaginative in his conception of how the brain works, in part because he lacked a good metaphor. Darwin believed that thoughts were secreted by the brain, just as digestive chemicals were. In his characteristically lyrical prose, he wrote: "Why is thought being a secretion of the brain more wonderful than gravity a property of matter?"[17] This was not an unreasonable hypothesis given that secretion had fundamental roles in the biology of Darwin's time. Why wouldn't thinking and volition work the same way? Nature certainly does reuse effective mechanistic strategies both within and across species. And neurotransmitter release, though it was unknown in Darwin's day, can be conceived as a secretion. But while Darwin's evolutionary theory shapes all of biology today, his conception of the brain (which may not have been meant to be taken seriously) is largely forgotten.

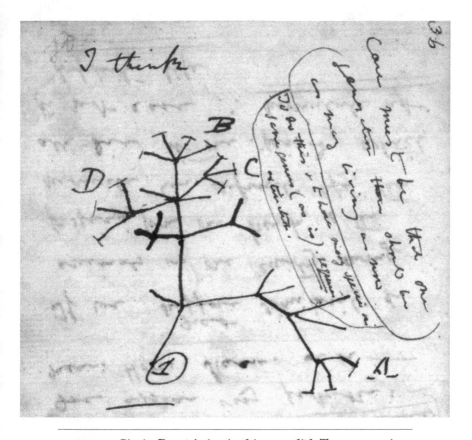

FIGURE 2.2 Charles Darwin's sketch of the tree of life. The tree metaphor helped Darwin organize his observations on heredity and variation into a coherent framework.

Image by Charles Darwin - Page 36 of Notebook B: [Transmutation of species (1837–1838)]. 'commenced… July 1837'; from Darwin Online, https://commons .wikimedia.org/w/index.php?curid=36638808.

Secretion theory is an example of a more literal approach to biological theory: a behavior observed in one system is appropriated to explain another system. In contrast, a metaphor can open new modes of conceiving the mechanics of a system beyond those already understood. Technology is an especially good target for brain metaphor. In part, this is because technology is cumulative, meaning that innovation builds on itself, just as in living things. More importantly, the brain, like technology, has a purpose—or many purposes. Metaphor provides not only a mechanism, but also a goal.

• • •

We can trace the evolution of brain-tech metaphors through the German philosopher Gottfried Leibniz, born in 1646, four years before Descartes died. The hydropower of Descartes's era had improved by Leibniz's time to the point where water could drive, not just fountains, but complex manufacturing operations such as textile factories. Leibniz rejected the vestiges of metaphysics in Descartes's worldview. The mind was not on a mystical plane, as Descartes thought, but rather was the result of brain operations. Leibniz wondered if these operations could be understood as mechanisms, and he suggested the metaphor of a mill. The mill metaphor suggests that the brain is a set of complex, coordinated mechanisms that are located next to one another. But the mill metaphor also makes a subtler point. It suggests that different processes may be operating at different levels in both mills and brains, and that the micro level did not necessarily tell us about the macro level. The idea is described by contemporary brain scientists Danielle Bassett and Michael Gazzaniga:[18]

> Emergence—of consciousness or otherwise—in the human brain can be thought of as characterizing the interaction between two broad levels: the mind and the physical brain. To visualize this dichotomy, imagine that you are walking with Leibniz through a mill. Consider that you can blow the mill up in size such that

all components are magnified and you can walk among them. All that you find are mechanical components that push against each other but there is little if any trace of the function of the whole mill represented at this level.

In other words, once technology reaches a certain level of complexity, the trick becomes the manner in which we combine many smaller tricks. Leibniz's mill metaphor captures this idea. The notion of nested, interlocking mechanisms is a prelude to complexity research of the twentieth century. In complex systems with many components, researchers often find that "more is different": new behavior emerges when a system has lots of elements, behavior that can't be predicted even with a full knowledge of how each element works.[19]

In the twentieth century, the mill metaphor for the brain morphed into the computer metaphor, partly by appropriating Leibniz's other innovations. Leibniz is best known for coinventing calculus with Isaac Newton. But he also built the most advanced mechanical calculating machine of his era (figure 2.3). Seeing the brain as a computer is a matter of combining these ideas: using the brain's complex millworks to perform systematic calculations.

The computer metaphor crept into neuroscience slowly. It entered via the transistor in the 1940s, as we will see. But from then until today, it has been the only game in town. The computer metaphor is worth stating clearly, as the philosopher Patricia Churchland and neuroscientist Terry Sejnowski did in 1992, in their landmark book *The Computational Brain*. They write: "Nervous systems and probably parts of nervous systems are themselves naturally evolved computers. . . . They represent features and relations in the world and they enable an animal to adapt to its circumstances."[20]

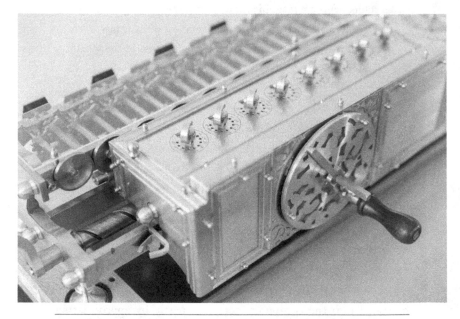

FIGURE 2.3 Replica of Leibniz's calculating machine circa 1700,
the stepped reckoner (from the Deutsches Museum, Munich).
This machine was the most sophisticated computer of its day.
Leibniz has been called the "patron saint" of cybernetics, a precursor
of contemporary computational neuroscience.
"Rechenmaschine_von_Leibniz_(Nachbau)" By Eremeev (CC-BY-SA 4.0)

There is no question that the computer metaphor provides major insights into the workings of brains. We will examine several examples of computer metaphor thinking in a critical way, but the usefulness of the metaphor should not be considered in doubt. We should acknowledge its importance and the thinkers who created it. This is especially important given that the computer metaphor was built on incomplete data and, in part, on incorrect assumptions regarding neurobiology.

A direct line can be traced between Leibniz and the computer metaphor for the brain. This connection was noted by a founder of the cybernetics movement, mathematician Norbert Wiener, who called Leibniz the movement's "patron saint."[21] Cybernetics initially aimed to understand network control and communication, especially as concerned the brain, but the practical implementation of these ideas mostly centered on computation. Today many identify cybernetics with computational neuroscience. From the beginning, a key idea in this field was to study the complex interactions in nervous systems in reference to computational circuits. As we will see, the theoretical progeny of cybernetics focused on machines that perform parallel computations and sequences of parallel computations, such as artificial neural networks. But it is worth remembering the emphasis on network communication and control in the original cybernetics approach.

Along with Wiener, neuropsychologist Warren McCulloch and computer scientist Walter Pitts helped found cybernetics with a seminal paper in 1943.[22] McCulloch and Pitts were interested in modeling how individual neurons work. They especially wanted to capture two properties: (1) neurons appeared to sum together excitatory signals received at the dendrites at a given moment; and (2) neurons produced spikes (action potentials) only when the cell's internal milieu reached a particular level of excitation in that moment.

To build a model, McCulloch and Pitts looked to the highest technology of the day: the electromechanical switch (and its cousin, the transistor, which was practically implemented a few years later). Switches made computers possible because they could be strung together to perform logical operations (if this and that are on, turn the next thing off). In other words, switches can perform computations. McCulloch and Pitts treated the inputs to a neuron's dendrites as switch-like elements: inputs were numerical values that could be positive or negative, corresponding to excitation (which makes the neuron more likely to fire) and inhibition (which makes the neuron less likely to fire). In their model, each input value is multiplied by a corresponding weight value, the weight being a kind of numerical bias in the switch. The resulting values are summed, and if the sum is greater than some small positive value, the neuron "fires" or produces a "1" as an output. If not, it remains in the "0" state and produces no spike.

The McCulloch and Pitts neuron was the first mathematical model of a neuron, and it has inspired countless studies and innovations, especially the invention of artificial neural networks. But today the model is useful mainly as an example of what is possible computationally using only neuron-like elements. It's not a useful model of how neurons actually fire, though it inspired successful later models of neurons that do predict firing fairly well.

McCulloch and Pitts showed that imagining a correspondence between the signal processing of individual neurons and switch arrays can give us new ways of thinking about whole brains. They didn't perform experiments to compare their model to real neurons. And in their time, spiking was well known to be more complicated than just the one-or-zero binary signal their model produced. Moreover, many assumptions on which the model is based are now known to be erroneous. Consider the first sentences of the 1943 paper that introduced their model: "The nervous system is a net of neurons, each having a soma and an axon. Their adjunctions, or synapses, are always between the axon of one neuron and the soma of another."[23] Today we know that several kinds of neurons lack axons, and that synapses can connect axons to dendrites, axons to axons, or dendrites to dendrites (and some neurons also possess direct, nonsynaptic connections called gap junctions that allow them to exchange chemicals). I highlight these faults not to impugn the work of McCulloch and Pitts but rather to show the power of a good if imperfect metaphor, and especially the power of a concrete technological metaphor that captures nature's tricks. The McCulloch-Pitts model highlights the fact that the way a neuron transforms a set of inputs into a set of outputs is something worthy of study. The system can be more easily recognized in reference to one humans have designed and are familiar with. Something like a computer.

McCulloch and Pitts didn't need to perform neurobiological experiments because what they were after was more akin to metaphor, or proto-theory. McCulloch would go on to write one of the clearest and earliest expressions of the computer metaphor, the 1949 essay "The Brain as a Computing Machine."[24] With this idea, computer-like neurons expanded to encompass processes that require many neurons in the brain, such as cognition. The McCulloch-Pitts model grew into what we now call artificial neural networks, or neural nets, which are used to model all manner of brain functions and are the basis of most AI.

The basic design of artificial neural networks stems from Frank Rosenblatt's modifications to the McCulloch-Pitts model in the late 1950s. Rosenblatt (figure 2.4), working at Cornell University, was interested in how a brain-like device could store information and make meaningful associations between chunks of sensory data. He built a machine he called the Perceptron.[25] It consisted of an (analog) electronic camera with an array of detectors that measured light intensity in a particular location—what we call today pixels. Each pixel was wired up in parallel to a summating detector that added up the input intensities of a number of pixels. Pixel intensity was scaled by a weight value associated with the resistance in the wire between a given pixel and a given detector. Weight values were initially random, but after each image was presented to the Perceptron, the system adjusted the set of weights according to a fixed rule. The rule one chooses—say, turn down all the weights when a letter E is present in the image, but not when an X is present—determines what the system will learn.

FIGURE 2.4 Frank Rosenblatt (1928–1971), inventor of the Perceptron, an early artificial neural network architecture.

Source: By Anonymous – http://www.peoples.ru/science/psihology /frank_rosenblatt/, CC BY-SA 4.0, https://commons.wikimedia.org/w/index .php?curid=64998425.

The main difference between the Perceptron and the McCulloch-Pitts neuron is the updating of weights. With this tweak, the machine could learn to identify simple pictures based solely on pixel values. In some of the earliest tests, Rosenblatt's machine could discriminate between black-and-white representations of the letters E and X rotated at various orientations with close to 100 percent success.

The device Rosenblatt built used electric motors to adjust potentiometers (which change electrical resistance) as a way to encode changes of weight values. Each of the hundreds of connections between input and output required a separate wire: in images of the machine the extraordinary rat's nest of wires is most striking (figure 2.5). Despite being an analog machine, the Perceptron performed essentially the same trick as the most sophisticated AI software today. McCulloch, Pitts, and Rosenblatt are together justly revered figures today, whose work, in addition to launching machine learning, helped found several disciplines in neuroscience, including cognitive science and psycholinguistics.

However, neural nets are a useful case study in how the computer metaphor aids but also limits our conception of brain function. Today, the dominance of the computer metaphor extends into almost all other corners of brain-related science, from psychiatry to social psychology. It is an unquestionably useful idea but one that cannot encompass all the brain's tricks.

Neural nets are built to represent things. Their job is to store and re-create information. The goal is to represent incoming data in such a way as to make a good decision or judgment. For example, a neural net can sort a set of images of either a car or an airplane into their respective category. The net does this by learning the kinds of fragmentary visual patterns that are common in pictures of cars and of airplanes, such as the characteristic curvatures of wheels or wings. In doing so, it represents the input images in the sense that it has taken the contours of the collected mass of pixel data. Neural nets don't care what the data signify; they only look for numerical patterns in the input. The goal of an artificial neural network is to arrive at an end state that captures some

FIGURE 2.5 The Mark I Perceptron that Rosenblatt and colleagues
constructed at the Cornell Aeronautical Laboratory in the late 1950s.
Source: (WP:NFCC#4), https://en.wikipedia.org/w/index.php?curid=47541432.

kinds of regularities in whatever streams of numbers are input to it as training data.

From the Perceptron, there grew a movement to expand and enhance artificial neural networks. After overcoming a period of skepticism and reappraisal in the 1960s, neural nets increased in sophistication and size, growing with increased hardware speed. Eventually they gave rise to the deep learning movement of today. Deep nets are a form of artificial neural networks that rose to prominence in the machine learning and computer vision communities in the 1990s and early 2000s.[26] They differ from the Perceptron mainly in terms of their sheer size, and especially the number of weights used to represent input data.[27]

Systems with just a single layer of neurons, like the first Perceptron, can perform basic distinctions between input data with some success, but they work much better if we add more layers. In particular, we connect every unit in the output layer with every unit in an additional output layer. In doing so, we create all-to-all connectivity between layers: each neuron in an additional layer has a weight value associated with every neuron in the preceding layer.[28] The original output layer is now called a hidden layer.

But distinguishing between cars and airplanes is a harder task than one might assume. The main challenge comes from the variability of lighting and geometry in an image. As described in chapter 1, the visual system must achieve invariance to irrelevant stimuli in order to perform tasks like recognition. For example, whether the image has the color red in it is not diagnostic of whether the image is a car or an airplane. Neural nets face the same challenge. To perform its task, a deep learning system needs to be invariant to types of variability that might be relevant in other contexts.

We feed images into the set of detectors pixel by pixel (figure 2.6). In a deep net, each input node (often called a "neuron") receives the pixel values for a particular neighborhood of

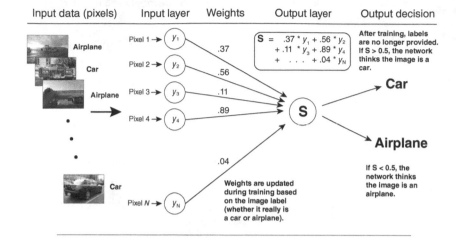

Input data (pixels)	Input layer	Weights	Output layer	Output decision

Airplane

Car

Airplane

•
•
•

Car

Pixel 1 → y_1

Pixel 2 → y_2

Pixel 3 → y_3

Pixel 4 → y_4

Pixel N → y_N

.37

.56

.11

.89

.04

$$S = .37 * y_1 + .56 * y_2 \\ + .11 * y_3 + .89 * y_4 \\ + \ldots + .04 * y_N$$

S

Car

Airplane

After training, labels are no longer provided. If S > 0.5, the network thinks the image is a car.

If S < 0.5, the network thinks the image is an airplane.

Weights are updated during training based on the image label (whether it really is a car or airplane).

FIGURE 2.6 Schematic design of an artificial neural network. The goal in this example is to generate a numerical value that indicates whether an image fed into the system contains a car or an airplane. Historically, this has been a challenging task for object recognition systems when images come "from the wild" and may omit portions of the object or be posed from noncanonical viewpoints. Artificial neural networks can solve this kind of problem with high accuracy after training. They do this by finding statistical regularities that differentiate the two classes of images. First, training images are fed in pixel by pixel to the input layer. Each node or "neuron" in the input layer has a weight value, which is multiplied by the input pixel value for a given training image—a car, for example. This product is summed across "neurons" and compared to a threshold value (0.5 in this case). If the sum is above this value, it means the system guessed correctly that the image was a car, and we move on to the next image. If the next image is also a car but it results in a sum less than 0.5, the weights are adjusted based on the size of this error. The inverse is done for airplane images. After training (i.e., "learning") over hundreds or thousands of images, weights are fixed and the system can be deployed on new images of cars and airplanes. In reality, more than one layer is usually needed, each with weights between successive input and output layers; systems with many layers are said to perform "deep learning." Image by Daniel Graham.

pixels in a given image. Each pixel within the detector's field of view is multiplied by a corresponding weight value stored in the input node, giving an output for each node in the layer. These outputs are treated as the inputs to the next layer "deeper" in the neural net.

In this second layer, the field of view of a node is expanded, and again each input datum is multiplied by the weight value corresponding to the link between the input node and output node. We repeat this a few more times. In later (deeper) layers, each node's field of view comprises the entire set of nodes in the preceding layer. These are referred to as "fully connected" layers since every input node connects to every preceding output node. Throughout the network, each connection has a corresponding weight.

After feeding in all the pixels in an image, we add up the outputs from the last layer, which gives us the network's guess as to whether the image is a car or an airplane. We also typically provide the neural net with a numeral indicating whether a car or airplane is actually present, say "0" for car, "1" for airplane. Then we compare the guess to the output we want, namely, the correct numeral for each image. If the difference between the summed output and the correct output is big, this means we need to make adjustments to the weights because the net has not learned how to make the decision very well. We adjust the entire set of weights by adding or subtracting a small value to them that is proportional to the wrongness of the last decision. If the difference between summed output and correct output is small, the net made a good guess, so the weights are not changed very much. Each weight value is thus shaped by pixel values across the set of images, as well as responses of other nodes. In this way, we can learn what is typical in a set of images of cars or airplanes in terms of weights. Deep nets are typically trained on half of the images—each of which has been labeled *car* or *airplane* by human viewers—and tested on the other half of the images that have no labels. Though a collection of weights is inscrutable to the human eye, it embodies the visual distinction between cars and airplanes (at least in the training set the net is given).

The advantage of having lots of weights in a network is that we can take the contours of the data in great detail.[29] Today, the best architectures can perform classifications of real-world images like handwritten numbers with accuracy close to 100 percent.[30] Wiring together even more layers is the basis of deep learning, and hence the accelerating number of weights in the system. In general, more hidden layers mean deeper, more fine-grained contour-taking. With more layers—and consequently many more weights—deep nets can distinguish, not just handwritten numbers, but the emotional connotations of an image, for example.[31]

Deep nets are especially popular today for modeling mechanisms of the brain's visual system, which we will focus on, though many other brain systems are studied with deep nets. In vision, this approach invokes elements of Marr's theory that vision is a kind of software that runs on the brain's hardware. A class of deep net models called convolutional neural networks (CNNs) are increasingly studied in relation to human vision. We will look at how CNNs work in the next chapter, but for now what's important is that deep CNNs have gotten good enough that they can predict how real visual neurons behave. CNN models have provided successful predictions in many domains beyond vision, too. Since we want theories in science that can predict things, deep learning models appear promising. If we can reliably predict how real neurons in living animals respond to pictures, maybe our models are doing the same thing the brain is.

Deep nets today are good enough that they can predict responses of neural populations in vision-related brain areas of the monkey based solely on the picture or sound the monkey is presented with. After training, they do this based solely on the images and sounds themselves. That is, changes in the electrical activity of a collection of neurons in awake animals can be predicted from pixel values or sound frequencies, using the machinery of deep learning. But before we rush to adopt deep learning as a guiding theory in neuroscience, we need to ask how good these predictions are.

To those outside neuroscience, this kind of experiment in monkeys may seem so sophisticated as to approach proverbial rocket science—it involves both brain surgery and AI. But the basic logic is not too complicated. We want to guess how a population of neurons sensitive to different regions of an image will respond to a given image, with only the image as a clue. It works like this. The scientist collects a few thousand digital photos of the natural world (perhaps gleaned from Google images). The monkey sees the images flash by for a few seconds each while its head is immobilized in a metal frame. Computers record a set of firing patterns produced by neurons numbering between a handful and, in the most ambitious studies, a few hundred or thousand cells. Most research of this kind involves surgically installed ports for electrodes, which are permanently affixed to the head. This way, a monkey can be tested numerous times.

The neurons we are talking about here are behind our ears in the outer sheet of the brain called the cortex. Specifically, they are in areas that analyze patterns of light and dark falling on our retina. These neurons are invariant to some changes in the position and size of light stimuli, and many seem to prefer certain objects such as faces. Scientists build a deep net that roughly approximates the sequential architecture of the visual system, with individual network units standing in for retinal and thalamic neurons and their connections. As we get to deeper layers, units learn about how local clusters of units lower down respond over time. This is where the "magic" happens, deep in the network, which is thought to model parts of the cortex involved primarily in high-level vision.

The network is trained on half of the image set shown to the monkey, so that the network's ultimate output—predicted number of spikes—comes to match the number of spikes when the animal is presented with the images. The network is now considered fully trained. Then it is deployed on the other half of the image set to see how well the prediction of firing matches the actual firing the deep net no longer has access to.

Here's the bad news. The best of these deep net systems predict less than half the variability in neuron activity over time.[32] Furthermore, because the system predicts average spike rates— but not the rarer bursts of high activity—it may be missing the most important signals. In particular, bursting behavior, a nearly ubiquitous form of neural activity in the brain, is not very well predicted by deep nets.[33]

It's important to say up front that predicting responses of neurons in awake monkeys in this part of the cortex is notoriously difficult, so any improvement in such methods should be welcomed and applauded. It is also not clear how much variability in a neuron's responses can ever be predicted since a given neuron in one individual's brain won't respond in exactly the same way as a corresponding neuron in a different individual (if indeed the idea of "corresponding neurons" makes any sense).

Predictions can generally generate four possible outcomes: hit, miss, false positive, or correct rejection. In contrast, typical deep net approaches measure variance accounted for, which indicates the degree to which variability in the data matches variability in the prediction.[34] Nor are the images the monkey sees very sensible: they show computer-generated objects such as a skinless,

FIGURE 2.7 Example stimuli from an experiment in which deep learning systems are trained to predict neural responses in the monkey brain when the animal is shown naturalistic images. The deep learning model aims to learn how a given neuron responds to images such as these.

Source: Daniel L. K. Yamins et al., "Performance-Optimized Hierarchical Models Predict Neural Responses in Higher Visual Cortex," *Proceedings of the National Academy of Sciences* 111, no. 23 (2014): 8619–8624. Images downloaded from https://github.com/dicarlolab/nrb.

manikin-like face, a surfboard, or a cow oriented in a random direction, floating in front of a random, unrelated natural landscape (see figure 2.7).

While many see deep learning as a tool that will revolutionize neuroscience, there are reasons for skepticism. Neuroscience may be placing too much faith in deep learning to solve the brain.[35] The limitations of deep learning are important because they impact our understanding of where to go next in neuroscience, and in particular how to study the networked structure of the brain. Deep learning is fully predicated on the computer metaphor, and the limitations of deep learning are in part limitations of the computer metaphor. As we will see, the internet metaphor makes very different and more plausible assumptions about brain network architecture compared to artificial neural networks and deep learning systems.

First of all, deep learning models are not necessarily any better as models than alternative non-deep models of neural spiking in the same parts of the visual cortex. Alternative models are more principled, in the sense that they posit rules of organization in relation to the specific kind of information coming into the system. For example, they may posit rules for detecting variations in face appearance and structure.[36] Deep learning, on the other hand, is a statistical picture. Constructing a deep learning model is something like building a clay model of a stone sculpture: we may create a decent copy of the sculptural forms, but we will learn little about how the original was made. And though most brain researchers disclaim deep nets as mechanistic models, the inescapable conclusion of this line of research is that the brain does use deep net–like mechanisms, since no other mechanism is hypothesized.

Deep nets invoke the idea of a network in their very name. But the deep net conception of networks is not the same as network scientists' conception of networks. Indeed, network scientists don't really consider artificial neural networks to be networks at all.[37] Instead, the power of deep nets comes from the brute force of connecting together many parallel computations in a strictly sequential "network" to perform adaptive computations. Unlike social networks—or brain networks—the network connectivity of deep nets is rather uninteresting. In figure 2.8, compare the network structure of the brain (*left*) to that of a deep net (*right*)— and to the internet (center). More to the point, the network architecture of deep nets is problematic for performing brain-like functions.

Deep nets almost always require all-to-all wiring between adjacent layers. This pattern of connectivity is needed every time we add another fully connected layer. In models of the monkey visual system, fully connected layers stand in for high-level

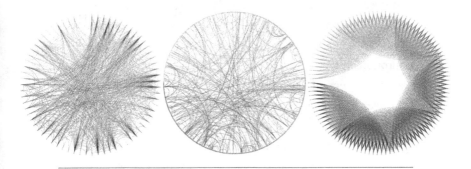

FIGURE 2.8 Comparison of network structure in the brain (part of the cortex of the macaque monkey (CoCoMac dataset), the internet (part of CAIDA-Skitter dataset), and a toy six-layer deep learning system. The brain is highly interconnected, with many local connections but also substantial numbers of shortcuts to more distant parts of the network. Any node is just a hop or two from almost any other node. The internet has a similar pattern, though nodes are less densely interconnected to each other compared to the brain. Deep nets, on the other hand, are rigidly sequential and require full interconnection between many layers. A given node is likely to be several hops from another node. Image by Daniel Graham.

processing in the brain's temporal lobe, where object categories may be computed. The problem is that nothing like full interconnectivity could exist in the human brain. It has been estimated that if all our neurons were fully interconnected, our head would need to be 20 kilometers in diameter to fit all the wiring.[38] Full interconnectivity only between adjacent areas, while less fantastical, is also not plausible. Connections between one cortical area and another are generally patchy or clustered, not uniformly distributed to all possible inputs in the neighboring brain part.[39] The developing brain, although more densely interconnected than the adult brain, is far from being fully connected.

Another problem with the network architecture of deep nets is that there are many hops between the beginning of processing and the end, that is, from the input layer to the output layer. Since units in a neural net are thought of as neurons, the hops correspond to synapses, or the gaps between neurons across which neurotransmitters travel. But the brain can't wait for a signal to traverse dozens of synapses to perform essential functions. For example, recognizing an object probably takes at most 150 milliseconds,[40] which is enough time for a signal to travel about five synapses.[41] Deep net models that suppose dozens of layers or operations to accomplish a task are implausible. Unfortunately, as machine learning researchers increasingly acknowledge current limitations in the performance of deep learning systems, their solution is often to add even more layers.

An artificial neural network is not a plausible model for communication because it was not designed for communication; it was designed for representation. Part of the problem is that we are thinking in terms of computation instead of communication.

Really, a deep net is just a simple computation repeated many times: it is a fancy name for combining matrices of numbers again and again. At heart, the mathematical mechanism of a neural network is matrix multiplication. This simply means multiplying elements of one array of numbers by corresponding elements in another array of numbers. The rest of a neural net is just concerned with updating numbers in the matrices being multiplied.

With enough numbers, we can describe an input space very well, and we can distinguish among relevant categories of images, as well as speech, financial market variations, and many other phenomena. Chess can be thought of as an input space with 64 dimensions, one for each square on the board. Each square potentially holds one of the 32 pieces. Changes over time to each square's state follow fixed rules. Any chess game can be thought of as a pattern in this space. With enough computing power to estimate how games will play out, a deep net can learn to crush any human opponent. Deep nets are unquestionably useful in these kinds of applications, when there are fixed rules and a relatively low number of dimensions. But they are not as useful in describing the flexible operation of biological brains with billions of interconnected neurons. Neural nets are an archetypal computer. But in being so good at computation, they cannot be expected to help us understand how real brains are able to communicate flexibly.

Artificial neural networks also learn in very different ways compared to brains. Deep learning is notoriously sensitive to the idiosyncrasies of the training set.[42] Brains are not like this at all. One of the main ways that all comparable brains are alike is in the way they learn: flexibility is required. Our brains do not require a specific training set. For example, each of us learns language from wildly variable training sets: from the beginning, the words we hear and the order we hear them in are never identical, and there are geographical differences in speech production. Yet all speakers of a language can communicate with one another from the moment they meet.

Language training of this kind is also mostly unlabeled. Most words we learn are not taught by pointing to an apple and saying "apple," then waiting for us to say "apple." Instead, they are learned through spontaneous mimicry and repeated association. Other species learn in much the same way. Birds, for example, learn their songs exclusively through untutored observation (in combination with genetic predispositions for song elements). Artificial neural networks, on the other hand, are initially random and typically require supervision in the form of labeled training (i.e., this is a car and this is an airplane).[43] Once the training set is learned, deep nets are poor at generalizing. For example, deep learning's promise of solving the invariance problem—the ability to recognize objects if their qualities are varied—has proven illusory so far.[44]

• • •

The dominant metaphors for the brain are powerful systems of thought and language that allow us to analyze our measurements in a coherent way, even when we are dealing with highly formalized experimental procedures and fine-grained data. But this power comes with a cost. It imposes a cognitive frame on our understanding of the brain, both in science and, as I will argue later, in society. We can disclaim the computer metaphor again and again, as many neuroscientists do, but unless we augment it with something else, we are liable to return to its logic, even unconsciously. The deep net movement, despite its success, is emblematic of our unspoken reliance on the computer metaphor.

As important as a new metaphor is, we need to be cautious about how it is deployed. Analogizing with contemporary high technology has its own particular risks due to our cognitive biases. Humans are unduly excited by hi-tech things like the internet, AI, and self-driving cars because the future seems limitless, hyperefficient, and inevitable. The possibilities for metaphors with high technology are likewise prone to overly rosy projections. We need to be vigilant against false or unrealistic hopes. We need to exploit what is useful in any metaphor and discard what is not.

In addition, not all advances in technology contribute useful metaphors. At the height of the Industrial Revolution, the parapsychologist Frederic Myers wrote that the brain was "a vast manufactory, in which thousands of looms, of complex and differing patterns, are habitually at work."[45] This metaphor, which builds on Leibniz's mill, was taken up by Sherrington, who wrote of the brain as an "enchanted loom where millions of flashing shuttles weave a dissolving pattern."[46] The image of the enchanted loom weaving our thoughts has endured. But even in Sherrington's day the metaphor was vague and intended largely as poetry (Sherrington proposed other, better metaphors, such as the valve, as well as the telecommunication switchboard, which we will discuss later in this chapter). There is some sense that the brain has many simultaneously operating machines making a beautiful pattern. And a loom evokes the growth patterns of axons and dendrites during brain development. But brains aren't primarily designed for manufacturing high volumes of identical things. The loom metaphor, however evocative as poetry, did not suggest theories of brain function as well as the computer metaphor eventually would.[47]

At this point you might say we don't need another bad analogy; we just need to look at the biology. An ion channel is not just a valve; it is also a dynamic biochemical system whose thermodynamics and genetic basis can be investigated. There is no question that basic biological processes can and should be investigated in the absence of metaphorical guidance. But an understanding of the complexity and interconnection of the larger system of a brain benefits from analogical thinking. As Descartes, Leibniz, and McCulloch and Pitts showed, a flawed analogy—even one based on incorrect assumptions—can spur new questions and lead to better scientific understanding, compared to an approach that supposes only an assemblage of specialized phenomena.

Alongside the computer in the history of metaphors for the brain, there has been a parallel but much less prominent strand of metaphorical thinking. It has been attached to the notion of the brain as a networked communication system. Some have traced this idea to Herbert Spencer, the polymath thinker of the second half of the nineteenth century who wrote of "numerous places of converging and diverging communication; each [able] to pass on in increased amounts the waves that distrurb them."[48] Spencer is derided by some as an "unread eminent Victorian with acute log-orrhea,"[49] and his writings about communication in neural networks are indeed vague. But Spencer's descriptions of networks of neural connections are supplemented by drawings of their connectivity. These images attest to Spencer's innovation in conceiving the nervous system as a complex network (see figure 2.9).

Spencer strongly influenced Sherrington and another great innovator in neuropsychology, Ivan Pavlov. Both won Nobel Prizes for Spencer-inspired research on the nervous system in the early twentieth century. Interestingly, both Pavlov and Sherrington also referred to the nervous system as a telecommunications switchboard, which was rising to ubiquity at the same time.

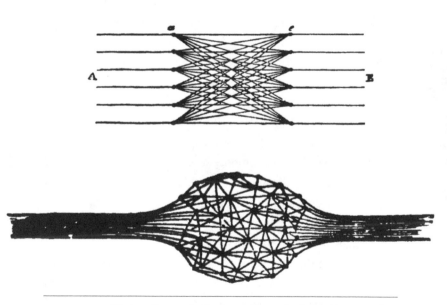

FIGURE 2.9 Herbert Spencer's diagrams of hypothetical brain network structures, 1896. Image source: https://babel.hathitrust.org/cgi/pt?id=mdp .39015058687925&view=1up&seq=553&size=175

While Pavlov wrote only briefly about the brain as a switch-board,[50] Sherrington is today associated with the notion that the central nervous system is like a telephone system (the loom meta-phor notwithstanding). He wrote that the brain is

> a vast network whose lines of conduction follow a certain scheme of pattern, but within that pattern the details of connection are, at the entrance to each common path, mutable. The gray matter [i.e., neuronal cell bodies] may be compared with a telephone exchange, where, from moment to moment, though the end-points of the system are fixed, the connections between starting points and terminal points are changed to suit passing requirements, as the functional points are shifted at a great railway junction. In order to realize the exchange at work, one must add to its purely spatial plan the temporal datum that within certain limits the connec-tions of the lines shift to and fro from minute to minute.[51]

What he is saying is that the brain must somehow allow for the flexible interchange of information, and it could do so in a similar way as a telephone or rail junction. The switchboard metaphor gained popularity and was for a time adopted in the popular press (see figure 2.10), but was soon to be overshadowed by the com-puter metaphor.

Sherrington seems to have imagined mechanical switches that change the connectivity or topology of the network over time. In order to connect one incoming line at a switchboard to another, both must first be disconnected from any other line. Once the two parties are connected, no other lines can access the chain of connections between the two parties. That is, the physical lay-out or topology of the network has changed. At a rail junction, changing the state of the switch likewise alters the topology of the rail network for as long as the switch stays in that state.

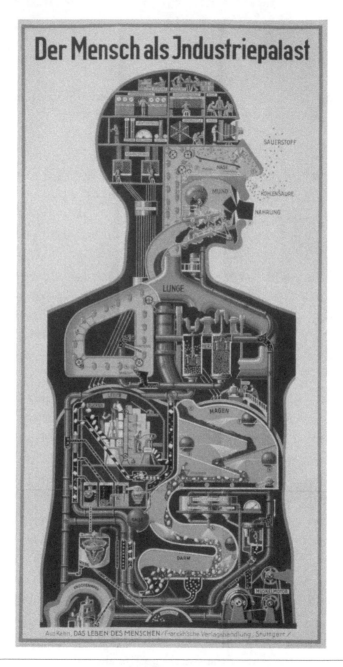

FIGURE 2.10 The brain as a switchboard, from a work of popular science by
Fritz Kahn, 1926. https://www.nlm.nih.gov/exhibition/dreamanatomy
/da_g_IV-A-oɪ.html.

It makes sense that Sherrington would look to these kinds of hi-tech systems to solve the problem of flexible routing. Aside from a power source, the critical technology in both telephone and rail systems is systematic routing. The whole purpose is for each node—a phone or station—to potentially be reachable from every other node on the network, and to plan for and deal with network traffic. But when network topology changes, usable paths between nodes change. Short paths may no longer be so short, and some are no longer viable at all.

We know that Sherrington's proposed mechanisms are not correct: changes in the topology of brain networks over short time periods—like the time needed to weigh evidence and make a decision—are negligible. Instead, we are learning that many potential mechanisms allow for the fast and flexible routing of signals without changes to network topology. But the metaphor of a switchboard made the problem visible and suggested a trick to solve it.

These ideas were a component of early cybernetics in the 1940s and 1950s. In practice, though, cybernetics researchers and those that followed almost always focused on computation at nodes in the network. What mattered was the network's output, rather than flexible communication across the whole network as envisioned by Sherrington, Pavlov, and indeed Spencer. But because they lacked an example of a sophisticated system for flexible, global intercommunication—in the 1940s, telephone networks ran on essentially the same routing protocol as telegraph networks of a century before—early cyberneticists can be forgiven for choosing computation as their metaphor instead.

• • •

Historically, scientists who have studied communication have also largely seen communication as computation, rather than as a way of flexibly exchanging information across a complex network. The mathematician Claude Shannon, inventor of information theory, was certainly concerned with communication, working as he did for Bell Telephone Laboratories. As part of a pioneering group of researchers who helped define the scope of early artificial intelligence, Shannon also had a deep interest in brains and cognition. In 1956, Shannon and colleagues envisioned a world where computers could understand language, learn about abstract concepts, and even be capable of creativity. But again, his focus—and that of almost everyone who followed—was on artificial neural network-like computations, rather than flexible network-wide communication.

Shannon's theory of information is often invoked in neuroscience but is of limited utility for understanding brain networks. Indeed, Shannon himself was skeptical about applying his theory of information outside of electronic signaling.[52] The framework of information theory deals almost exclusively with measuring communication across a single link, like a wire between sender and receiver. For this, it is a good theory. It can tell us, for example, how thick a wire we need to achieve decent audio quality in a phone conversation.

Information theory is good for making communication across a given wire efficient. Let's say we have a message we want to send across this wire to our friend. Pretend, again, that our friend is expecting a message from us about when we are meeting for a concert that evening. What we are really trying to do with our message is to reduce our friend's uncertainty about when to meet. We could probably get away with saying, simply, "7" since most concerts are in the evening. But to be sure, we might instead write "7p." If this is all we want to communicate, there is no reason to include any extra characters to write "p.m." Adding .m. would be redundant, and sending these characters across the wire would be wasteful. Ideally, all communications traveling down the wire should be those that are only necessary for the receiver to know what we want to say. In a sense, we want our message to be as close to gibberish as possible while still passing all of the intended information. Mathematically, gibberish means we want to use all the symbols in our alphabet the same number of times over the course of many messages. When some symbols such as letters are used more than others—or only adjacent to certain other letters, like "p.m."—we have redundancy. Redundancy is wasteful to communicate, since it contains information that can be inferred by the receiver. We can build cheaper links of the same quality if we remove these kinds of redundancy.

The mathematical machinery of information theory tells us very precisely how efficient our code is—be it an alphabet, a dictionary, a set of electronic signals, or any other code system—relative to an ideal system. The infrastructure of the internet makes use of Shannon's theory of information every day in how it passes messages from node to node. But Shannon's theory is silent on how to move information across more than one hop in a network, how to control the real-time flow of signals, how to correct errors, and how to achieve global flexibility.

Again, this is because it was conceived largely as a computational theory, and has in turn been applied to brains as such. There has been some progress toward fundamental Shannonesque laws of communication across complex networks.[53] But unlike Shannon's theory, which has been widely adopted in the study of neural coding, newer information theoretical frameworks for complex networks are incomplete and have yet to be widely applied to the brain. With a shift of metaphor, we might be motivated to think more about this problem.

A key issue in applying Shannon's theory of information to the brain is that the theory deals only with the probabilities of symbol usage, not the meaning of the symbols. This, in turn, is another reason that the theory is of limited utility in brains, even when we are considering signals in single neurons. As we will see in the next chapter, we don't know what constitutes a "symbol" within the brain's internal codebook. But using principles from information theory, we can still gain insights into how the brain manages information, especially if we try to understand how the brain as a whole operates in its natural environment. What is lacking is in part a frame and a language for discussing efficient, reliable communication among large numbers of neural components.

In some ways, it is odd that it has taken so long to recognize that flexible, efficient, reliable communication is precisely what both the internet and the brain do. The key innovations—such as a system for dividing up messages into chunks of fixed size—have been known for more than half a century. We should not expect that the brain works exactly like the internet. But tricks similar to those used by the internet are necessary in the brain as well.

Ultimately, we still need metaphors for the brain because we are so far away from understanding the brain. We also still need metaphors because the brain does many different things. But even if we wanted to dispense with metaphor we couldn't, because the brain is each one of us. We identify very strongly with it, and we need a way of understanding it. Currently, scientists and nonscientists alike conceive of their brains as computer-like entities in their day-to-day lives. We easily extrapolate from this basic metaphor to explain our individual traits and experiences. Are we really *hardwired* for motherhood or mathematics? Have we really been *running on autopilot*, or are we really *defragging* after a long day at work? It feels so natural to make these comparisons because we are all confined within the computer metaphor. Brain science is likewise limited. But to escape, we first need to understand in broad terms what is known and what is unknown about brains and how they process information.

3

WHAT WE DON'T KNOW
ABOUT BRAINS

T'S easy to get the impression that science has by now achieved a very good understanding of how brains work. The onslaught of popular articles and books that aim to explain the workings of our brain, as well as the proliferation of the prefix *neuro* in other fields from economics to marketing to art history, can give the impression that neuroscience is a solved problem and that the answers neuroscientists have discovered can guide other fields. There is even an expectation that our current basis for understanding the brain will in the next few years bring computer-based brain enhancement. If this is the case, new theoretical frameworks are unnecessary, since the current one has succeeded.

By framing this chapter around what we don't know, I am by no means trying to diminish the accomplishments of outstanding neuroscientists working today and in the past who were guided by the computer metaphor (or any other metaphor). Major progress has indeed been made in understanding brains, especially recently, and much of it by following the computer metaphor.

Progress has come in part because the myriad biochemical and biophysical processes that occur within single neurons can be studied with relative ease: individual cells can be coaxed to grow outside of living animals. Gene editing in mice and rats can also produce individuals who lack specific neuron types, or have other fundamental modifications in their physiological functions, as well as cell lines with particular characteristics. This can help us understand what a particular neuron does and how it computes.

But whole brains are not understood to a comparable degree. Basic descriptive quantities are not known with reasonable accuracy, least of all for humans. We do know that the total number of neurons in human brains is 86 billion or so, but this was only accurately measured around 2009.[1] A parts list for the cortex of a mouse brain, comprising 140 cell types, was only determined in 2018.[2] But this list, for an animal separated from us by more than 80 million years of evolutionary divergence, may still be partial, and it leaves out a constellation of other neural systems.

Some basic questions haven't been studied in a concerted way. How many neurons is a given neuron connected to? The canonical estimate is that this number is around 10,000.[3] But no one really knows. How many synapses are there in the brain as a whole? The number is guesstimated in the hundreds of trillions, but again no one has counted them or estimated their number in a systematic way.[4]

Our ignorance extends to basic aspects of neural activity on whole-brain networks.[5] How many neurons are active at the same time, in a given region or across the whole brain? Recall that only 10 percent of our brain can be very active at a given time because firing is so costly for the body's metabolism. And at most, individual neurons can fire during 10 percent of their lifetimes. But the 10 percent figure is a very rough estimate.[6] No one has any idea how much variability there is in collective firing. The number of neurons that fire at once varies in ways we can't summarize— or even measure. This is true both within brain regions and across the brain as a whole.

One quantity we would especially like to know with precision is how many hops or synapses separate a typical neuron from any other neuron in the brain. The value is small, probably around 3 or 4 for most neurons, and this estimate is a key consideration in our application of the internet metaphor. But though many researchers are providing data that increasingly help to achieve a precise value, the estimate is still rough.

There are a number of reasons why our understanding of brains is so limited. Many of the limitations are procedural: it boils down to the fact that living brains, particularly human brains, are difficult to study, whereas single neurons are tractable. But another factor may be neuroscience's theoretical bias toward single neurons, which is entwined with the computer metaphor.

There are alternative approaches to understanding the complexity of brains, especially those that emphasize holism. Though we don't understand overall brain activity very well at present, we may start to understand the brain from a holistic perspective if we think of it as an evolved, internet-like network.

We know brains are made of neurons, along with helper cells called glia. Neurons are just a kind of cell, one that performs the same basic processes shared with muscle cells, kidney cells, or blood cells. Like these other kinds of cells, neurons produce energy, read out genes, repair the cell, remove waste, and so on. Really, the specialized function of most cells—say, movement for a muscle cell—is overshadowed by the many shared metabolic processes that keep all cells alive and healthy. Thus insights gained in other areas of biology related to basic cellular metabolisms have advanced our understanding of neurons. But what we really want to know about is neurons' specialization for information processing and signaling.

If we consider only a single neuron in a petri dish, we know a tremendous amount, from miniscule biochemical components to the behavior of the whole cell. We understand the basic genetic and developmental processes of many neuron types in great detail. Cellular processes specific to neurons are also well understood, such as the bundling up and delivery of neurotransmitters to the synapse, thanks to especially intense investigation of this area in the early twenty-first century.

Neurons send signals, but in ways that go well beyond the signaling occurring in other kinds of cells. All cells in the body send signals to one another, and many also receive signals from the environment. These signals might carry a message from a secretory gland to cells with certain hormone receptors telling those cells to produce a particular biochemical.

But no biological system besides the nervous system is designed to deal in pure information. In other cell-signaling processes, the message carrier changes utterly after each transmission, from one chemical to another, and a single chemical reception can trigger an avalanche of other chemicals to be released. But in the brain, a message can traverse many hops seemingly without changing its basic currency: spurts of electrical excitation.

Neurons are designed to communicate with one another, and this is what differentiates neurons from other kinds of cells. Communication is the design goal of neurons, just as the ability to do mechanical work is the design goal of muscle cells.[7] The main way neurons exchange messages is via synapses, which are very small gaps between neurons. Surrounding neurons, and in the gaps between them called synapses, is a liquid environment containing various dissolved chemicals (see figure 3.1). The synapse is not empty or fully isolated, so right away, the reliability of communication between neurons becomes an issue. Getting an electrical signal across the wet synaptic gap is not so easy, since liquids don't conduct electrical signals as well as the metal in a wire does.

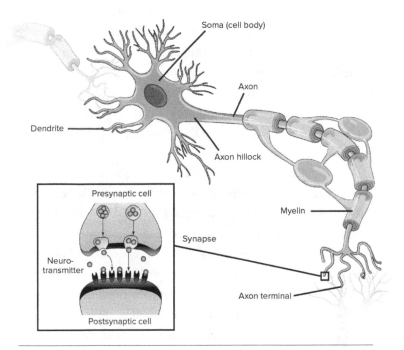

FIGURE 3.1 A neuron is a cell that is specialized for communication. It receives chemical signals (neurotransmitters) at its dendrites, which are summed together. If enough stimulation arrives in a short period, the neuron produces an electrical excitation that travels down the axon. Myelin sheathing helps insulate the axon, allowing the electrical signal to travel more efficiently. When the electrical signal arrives at the axon terminal, this triggers the release of neurotransmitters, which travel across the synapse to the next neuron in the network. Image modified from "Neurons and Glial Cells: Figure 2" and "Synapse," by OpenStax College, Biology (CC BY 3.0).

https://www.khanacademy.org/science/biology/human-biology/neuron-nervous-system/a/overview-of-neuron-structure-and-function.

Neurons translate spikes into pulses of neurotransmitters that are expelled into the synapse. The pulsed nature of signals is a strategy shared by all nervous systems as a way to make sure that those listening—neurons on the other side of the synapse—get the message. The specific solution nature has discovered is to use spikes. A spike from a given neuron means it produces a fast, stereotyped change in voltage over time. Each time that neuron fires, the spike's voltage change pattern is pretty much the same. Spikes initiate neurotransmitter release at the end of the axon (called the axon terminal). Crucially, spikes are padded in time by silence to make sure neurons listening on the other side of the synapse can hear the message loud and clear.

Neurons have input zones and output zones. Inputs (such as neurotransmitters) arrive at the dendrites and cell body (the bulge in the cell where the nucleus lives) and exit via the terminals of the axon (releasing more neurotransmitters). Spikes essentially summarize the signals' input to a neuron and pass this summary along to all other cells that are listening at the end of the axon, much as McCulloch and Pitts described them. Within a small window of time, in the range of milliseconds, the neuron has a chance to produce a spike if the incoming signals generate enough excitation. Neurotransmitter molecules passed across the synapse provide the excitation. Neurotransmitters can also negate, through inhibition, the overall excitation delivered across the synapse. If the total level of input excitation minus input inhibition is sufficiently high for a given neuron in a given instant, the neuron will produce its own spike. The spike travels from the input zone (dendrites and cell body) to the output zone (axon terminals) via the axon. When the spike reaches the output zone, it triggers the release of neurotransmitters, which are prepackaged just inside the axon terminal, standing ready to be released. Once in the synapse, neurotransmitters drift to the dendrites of the next neuron and begin the process again (spikes passed via neurotransmitters to muscle cells initiate action).

Given that there are around 86 billion neurons in our heads (along with hundreds of billions of glial cells), how did the study of brains come to focus so relentlessly on the single neuron, and on the computations single neurons perform? Part of the problem is that recording from more than a few neurons at a time is very difficult. Our attention naturally fixes on the few cells we can study at a time—or just one cell. Today, even the most ambitious and well-funded studies can accurately measure activity in only a few hundred neurons at a time, or less than a ten-millionth of the human brain as a whole. This is a major technical challenge. Of course, there are established technologies like magnetic resonance brain imaging and newer ones such as calcium imaging that attempt to measure large-scale activity in brains. These approaches have some utility, but they have also been called a "fantasy" by some neuroscientists.[8] Although attractive, imaging approaches unfortunately lack the resolution in space and time to accurately measure real interactions between individual neurons. Without good measurements of activity on brain networks, researchers have focused instead on understanding electrical responses in single neurons, and they have largely succeeded.

But this does not mean we understand brains. Aside from procedural limitations, there are biases in the dominant theoretical program of brain science that have also locked us into a relentless focus on single neurons and their computations.

The origins of the theoretical focus on single neurons lie in the founding debates of modern neuroscience. In the late nineteenth century, Camillo Golgi invented a technique for highlighting individual cells to reveal their intricate structure. Golgi's cell-staining technique, using solutions of dissolved silver, was applied to brain tissue. Without cell staining, the gross anatomy and fine structure of the brain do not give much hint of the brain's interconnectivity—it just looks like a rubbery blob with faint differences in coloration. The true structure was revealed with staining at the microscopic level. What Golgi and others saw in the microscope when they stained slices of brain tissue was an intricate net of fine filaments extending in graceful arcs and branches.

In the early twentieth century, anatomists and physiologists studying the brain could not agree whether brains were a unified tangle like a fishing net or an agglomeration of individuated units. The first camp, which included Golgi, were referred to as reticular theorists (*reticulum* is Latin for "little net"). They pointed to the apparent lack of a gap between one filament tangle and another.

In opposition, those subscribing to what became known as the neuron doctrine, such as Charles Sherrington and Santiago Ramón y Cajal, argued that the tangles had well-defined subunits—what we now call neurons. They pointed to the fact that cutting a part of the net led to local atrophy, which seemed to be limited only to the damaged subunit. Supporters of the neuron doctrine also noted that individual brain units seem to develop from distinct precursor units.[9] More generally, scientists judged that cleaving the problem of how brains work at the joints, so to speak, was the only way forward. So they focused intensely on single neurons. Over time, reticular theory became untenable in part because it made brain tissue difficult to study. If everything is part of one enormous and amazingly complex thing, how do we even begin to study it?

Guided by the neuron doctrine, researchers in the early twentieth century began with trying to understand how neurons pass signals from one cell to the next. Because they couldn't see the tiny synapse, they had to approach the problem indirectly (it was only in the 1950s that the synapse could be observed directly using electron microscopes).

Two theories emerged as to how signals were passed between cells: chemical transmission and electrical transmission. Chemical transmission means passing signals using the language of chemical concentrations: how much of a given molecule is in a given space at a given time. Electrical transmission means an exchange of electrons between neurons (electrons being bound to atomic nuclei, together making charged ions). While strong proof in favor of one or the other theory of transmission was absent, most agreed that there was some kind of transmission across the gap, though some retained faith in reticular theory, or were at least agnostic about the autonomy of neurons.

After a magnificent experiment by Otto Loewi in 1921, few could cling to reticular theory. Working in Vienna, Loewi showed that sufficient concentrations of chemicals could indeed transmit neural information from cell to cell. Loewi demonstrated this in an experiment that stands today among the most elegant in all of biology and indeed all of science. It involved the neural system for controlling heart rate. A beating heart was removed from a frog and isolated in nutritive fluid. Loewi electrically stimulated the vagus nerve that innervates the heart, which was known to cause the beating to slow. Then he transferred the fluid surrounding the heart into a vessel containing a second beating frog heart. Remarkably, this fluid alone caused the second heart's beating to slow also. The same effect was observed when the accelerator nerve of the first heart was stimulated, increasing the heart rate in the first heart, and the surrounding fluid again transferred to the second heart. In this case, the second heart's beating increased. See figure 3.2.

Loewi's experiment showed that some chemical was emitted during the process of increasing or decreasing the heartbeat through stimulation of neurons in the nerve. This chemical was responsible for transmitting the neural information to muscle cells. Electrical currents did not directly pass from the nerve to the heart muscle cells, since an electrical current from the first heart would quickly dissipate in the fluid and therefore would not transfer to the vessel with the second heart. Instead, stimulating

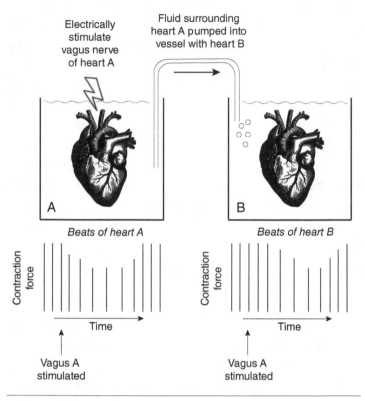

FIGURE 3.2 Otto Loewi showed that neural activity results in the release of chemicals that cause action in heart muscles. He stimulated the vagus nerve of an isolated frog heart, which caused the heart's rhythm to slow and weaken, and he collected fluid from this vessel immediately afterward. When transferred to a vessel with a second (unstimulated) heart, the fluid alone caused the second heart's rhythm to slow and weaken as well.

The chemicals were later identified as neurotransmitters; they allow communication not just from neurons to muscle but also between neurons. Image by Daniel Graham.

the neurons in the vagus nerve caused something to leak out of them, presumably at the point where they met other cells. The chemicals in the fluid that acted on the heart muscle cells were later identified as neurotransmitters. They deliver signals both from neurons to muscle cells, as in the Loewi experiment, and from neurons to other neurons, which is what we are most interested in.

Loewi's experiment did not put to rest the question of whether neurons communicate with each other via electrical or chemical means; today we know they can use both mechanisms. But from this point forward, the neuron doctrine has been dominant. Loewi showed decisively that neurons were not directly connected, since one could collect the action-producing chemicals that leaked out of them. The neuron doctrine won the day. Soon, it would seamlessly integrate with the computer metaphor.

In the 1930s and 1940s, scientists focused increasing attention on the pattern of excitation during a spike. By recording currents generated by single neurons (which is easiest in very large neurons such as one found in the squid), they gathered a wealth of new electrophysiological data. With this basis of knowledge, biophysicists Alan Hodgkin and Andrew Huxley described the process mathematically in 1952. Their goal was to approximate the physical forces—electromagnetic and mechanical—that animate a prototypical neuron. Their model used the language of differential equations to improve upon the McCulloch-Pitts model. As such, it implicitly adopted the computer metaphor: the goal was to describe a computed output given a particular input.

The Hodgkin-Huxley model treats a neuron as a tube of liquid with ions dissolved in it. The ions leak out of the neuron into the liquid around the cell at a set rate, reducing the voltage of the neuron relative to its environment. At the same time, excitations that raise the voltage of the cell come into the dendrites at frequent intervals. By writing mathematical descriptors of these rates of change and connecting them through fixed rules using calculus, the Hodgkin-Huxley model can predict how much excitation needs to be applied to the dendrites to generate a spike during a given amount of time.

With this model, we can predict how an isolated neuron is likely to respond to just about any incoming pattern of excitation. In other words, we can model the computations the neuron performs on its inputs. For a neuron at rest, a given input always leads to the same output. By changing only a few parameters, such as the size of the axon and the speed that ions leak, descriptions like the Hodgkin-Huxley model can predict the basic spiking behavior of most neurons in the brain fairly well using just a few free parameters. The Hodgkin-Huxley picture is useful for different classes of neurons (those differing in shape and other characteristics), for neurons located in different brain systems (from the olfactory system to the brainstem), and for neurons in essentially all animal species (and even related plant cells). Relatively simple spiking neuron models based on this model now succeed in predicting spiking responses of neurons in petri dishes to arbitrary patterns of electrical zaps with over 90 percent accuracy.[10]

The Hodgkin-Huxley model is not perfect. It doesn't capture detailed neuron activity such as current changes in between spikes. Moreover, things get complicated when the tube of liquid we are considering has many branches, like those in most dendritic trees.[11] Axons are also considerably more complicated than previously believed; they are not just long, uniform tubes. We know today that axons interact with other axons, primarily by exchanging ions directly. Sometimes a patch of axon can generate a spike without any participation of that neuron's dendrites or cell bodies, but this leads to neurotransmitter release at the neuron's axon terminals all the same. Conversely, sometimes a spike arrives at the end of the axon but fails to trigger neurotransmitter release. The spike can even go backwards toward the cell body. All of these interactions can affect classical spiking initiated by the dendrites over subsequent minutes. Recent evidence from living mammals suggests that a substantial proportion of spikes—

30 percent in a memory area called the hippocampus—are produced by axon-axon interactions rather than by axon-to-dendrite excitation, as would be supposed by the neuron doctrine.[12] None of these effects are predicted by the Hodgkin-Huxley model, or by those built on it.

But as a theoretical tool, the Hodgkin-Huxley picture captures the essential trick involved in making a spike, and as such is a remarkable achievement (which won Hodgkin and Huxley the Nobel Prize in physiology or medicine in 1963). The undeniable success of the Hodgkin-Huxley model and its progeny today is a direct result of more than a century of emphasis on the single neuron in research on brains, which has for much of this time been guided by the computer metaphor.

Focusing on the computations of neurons in isolation has even allowed us to study the apparent meaning of spikes. Once the mechanism of transmission was discovered, the task became one of understanding what one neuron is trying to tell another neuron in the language of spikes. This remains the great challenge of neuroscience today: the decipherment of the neural code. Just as archaeologists work out the meaning of an unknown script based on surviving fragments of text, neuroscientists try to understand the language of neurons based on samples of activity in the brain's coding units.

Major successes have come in the decipherment of the neural code in the brain's visual centers. The computations produced by neurons in these areas show clever strategies for measuring visual qualities of the world. But this approach can obscure the networked nature of the visual system and indeed the network architecture of the whole brain.

A special type of visual neuron was discovered by David Hubel and Torsten Wiesel, the Nobel Prize winners we met in chapter 1.[13] Hubel and Wiesel were interested in understanding the patterns of light that would make a neuron in the rear of the brain produce lots of spikes. The assumption was that visual information captured by the eye needs to be taken apart by local analyzers operating in the brain. It was known at the time that the first few layers of the network of the visual system—in the retina, at the back of the eye, and in the thalamus—resemble a "feed-forward" system: neurons are arranged in sequence to respond to increasingly complex visual features.

Mainly from work in frog eyes, it was known that the system starts by measuring light in the smallest dots and then compares collections of dots to a surrounding circular area on the retina. In addition, it had been known for fifty years that a particular rear part of the brain—an area called V1, or primary visual cortex—has a map of the visual world. Studies of gunshot victims whose visual cortex was damaged (casualties of the Russo-Japanese War and the First World War) showed that each part of the retina corresponds to a particular area of territory in V1. In terms of neural coding, although single neurons in V1 show little or no response to a uniform field of light, larger populations of neurons respond roughly in proportion to the amount of light hitting the corresponding array of photoreceptors in the retina.[14] So to a first approximation, V1 activity comprises a map of the brightest and darkest parts of an image.[15]

But Hubel and Wiesel's initial observations in the late 1950s suggested that the responses of single neurons to various patterns of light were more complicated than just more light equals more spikes. They figured that these neurons, like those in the retina, respond to particular shapes or patterns of light. But did the dots and disks detected in the retina just get bigger? Did they form a kind of hexagonal grid? Did they care about motion or color? Hubel and Wiesel had little idea what to expect as to the cells' pattern preferences.

What they found, when they measured electrical pulses in V1 of a cat, was that many cells responded best to rectangle-shaped bars of light. But the cells did so in interesting ways. What's happening here is that cells that are earlier in the visual system pool their responses together: some cells' responses are added together, making more pulses, while others cancel out. The net effect is that later cells can encode more complex spatial patterns. By firing when a particular pattern is present on the retina, the system can encrypt or encode the visual world. The visual world needs encrypting because object identity and other features like color or motion are not inherent features of the physical world. Rather, they are interpretations of the reflected light in the visual world that are useful to us for one reason or another.

Cats lack one of the three cone photoreceptors we have but otherwise have a similar organization of the early visual system. The cats tested by Hubel and Wiesel were anesthetized and had their eye propped open and pointed toward a projector screen in a darkened room. Hubel and Wiesel shined a variety of bright patterns like dots of light at the screen using a slide projector. They measured changes in voltage in the visual cortex using electrodes that penetrated the brain. Responses were recorded and simultaneously turned into sound.

First, they had to listen for the brief popcorn-popping sound of rapid spiking that would occasionally happen as they swept the dots across the screen. A given neuron would only fire when the spot was in the right location on the screen. This would tell them which part of visual space a given cell is sensitive to. Each neuron is fixed in place so it always responds to essentially the same portion of retinal space (and to the same part of world space if the head and eyes are immobilized, as in this experiment; in natural vision, as the eye moves, new patterns of light are constantly beamed to the cell).

After determining which part of visual space a cell responded to at least a little bit, Hubel and Wiesel had to find the best pattern of light for making that cell produce a lot of spikes. They stumbled on the most effective pattern—glowing rectangles—by accident. Stimulus patterns such as white circles of different sizes on a black background were printed on glass slides and moved around the projector screen to find a given cell's area of sensitivity. These didn't produce much response. Hubel and Wiesel hadn't initially made slides with rectangular patterns. But they noticed that cells would sometimes fire vigorously when they took slides out of the projector—precisely when the straight dark edge of the slide passed across a neuron's field of view. They soon found that each cell preferred bars of light at a particular orientation and location.

Across dozens of cells, the entire 180 degrees of orientation was captured by one cell or another, along with variations in width. Cells that behaved in this way were termed simple cells. The computation performed by simple cells was to add together stimulation in the bright part of the rectangular stimulus and to subtract stimulation in adjacent rectangular areas where there was darkness. When this happened, the cell encoded a corresponding pattern in the world.

Other cells in V1 responded to a bar of light at a particular orientation and location surrounded by darkness, but they would also respond to bars that were parallel to that orientation and shone at a different location nearby (as long as they were surrounded by darkness). These cells were called complex cells. Their computation was to add together light in a rectangle of a particular orientation and to be invariant to local position.

It may have been enough to discover and explain the computations occurring in simple and complex cells (along with other kinds of neurons in cat V1) in terms of the geometry of the visual world, and to characterize them in great detail, to justify the Nobel Prize for Hubel and Wiesel. But they went further by proposing a clever and plausible explanation for how V1 cells were connected in a network.

Hubel and Wiesel proposed that complex cells listen to a pool of appropriately chosen simple cells. In this view, a complex cell becomes a kind of OR computation, with high activity in any of its inputs being sufficient to make the complex cell fire.[16] See figure 3.3. The idea was that each local neighborhood in visual space presumably is served by numerous simple cells of the same orientation preference, located at different points in that neighborhood. If a complex cell listened only to those simple cells that preferred a particular orientation and that were "watching" contiguous but distinct regions of visual space, the complex cell would fire when any one of those simple cells fired. Thus, a particular rectangle-like pattern in the world would be detected even if its position varied somewhat. Hubel and Wiesel proposed that the visual system uses complex cells to achieve invariance to position, while retaining selectivity for a particular orientation.

Invariance is critical to visual function. The brain is constantly faced with the problem of picking out a particular pattern but remaining insensitive to certain variations in that pattern, such as position. We need to be able to recognize our favorite coffee mug no matter where on our retina its image happens to land. But a mug is not much more than a collection of glowing rectangles. Simple cells break any scene into pieces and encode it as, essentially, rectangles. To help us match the mug-like thing in our field of view to the stored representation of the object, "my mug," we need invariance. The variation to which complex cells are insensitive is a change in position of a bar of light (within a limited window). With a collection of complex cells tuned to a collection of orientations that define a mug, an appropriately wired set of

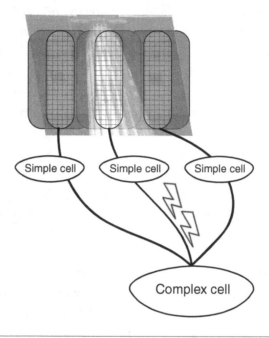

FIGURE 3.3 A hypothetical wiring diagram of a complex cell in the primary visual cortex, proposed by David Hubel and Torsten Wiesel. Wiring is such that high activity in any of the three simple cells will cause the complex cell to fire. Here, the simple cells are active when the eyes are presented with a bright vertical patch bordered by dark areas. The column image matches this pattern well and would make this cell fire vigorously (though in reality simple cells respond to much smaller areas of space). The complex cell's wiring pattern allows it to respond to the vertical pattern across a wide expanse of visual space, thereby achieving invariance to shift. However, this network architecture has not been directly observed. See David H. Hubel and Torsten N. Wiesel. "Receptive Fields, Binocular Interaction and Functional Architecture in the Cat's Visual Cortex." *Journal of Physiology* 160, no. 1 (1962): 106–154. Image by Daniel Graham.

higher-level detectors could use complex cell responses to find the pattern wherever it happens to be in our field of view.

Hubel and Wiesel's network model was a potential mechanism for invariance. It explained the data well and seemed to solve a major problem in vision. Here was a triumph of the neuron doctrine, augmented by the computer metaphor: Hubel and Wiesel achieved an understanding of not only how, but also why, a particular kind of neuron behaves as it does. It was a major step in deciphering the neural code.

It's not always recognized today that Hubel and Wiesel's landmark work (elaborated by many others) offers us, ultimately, an explanation about what is connected to what in the brain. It is a statement about the network architecture of the brain: how populations of simple cells are wired up to populations of complex cells. The problem, however, is that Hubel and Wiesel did not observe any of the proposed patterns of wiring of simple and complex cells, and no one has since.[17]

The reason that no one has confirmed this wiring pattern is again largely one of procedural limitations. To determine whether a cell is a simple or complex cell (or some other kind), and what sorts of patterns it responds to, we need a living neuron in a living animal.[18] A cell that responds like a complex cell cannot be grown in the lab. Its complex cell identity is fully determined by the pattern of connections from the eye to the cortex (and within the cortex). But to know which other cells a given cell is connected to, we need a dead animal, with its brain sliced into thin pieces. Determining what a cell does, and how it is wired—in the same experiment—remains an unsolved problem.[19]

Thus, there are major procedural obstacles in studying network communication in brains. We would like to know, for example, when a given neuron spikes, what is the trace of all the activity this action causes elsewhere on the network. At present, we have no way of measuring this. But at least part of the problem is our theoretical frame of the computer metaphor.

Even though we know pretty well how individual neurons work, we can't just treat the brain as an agglomeration of isolated neurons operating in sequence. We are dealing with a multiplicity of heterogeneous elements in neuroscience. This is a different situation compared to the physical sciences, where great simplifications are possible by treating every electron or every carbon atom as identical. We don't have that in the brain. Neurons of a given kind are not alike, and they will respond in different ways at different points in their life and based on their position in the brain. Nor is each spike alike. Many neurons hardly spike at all, as we will see. We also now know that spiking neurons send meaningful and reliable signals to other neurons without spiking.

But computational models of the brain today assume that neurons are much like identical electrons or carbon atoms and that whole brains can be understood as collections of identical spiking neurons. In this view, brains can be understood as collections of "point neurons" that behave basically as Hodgkin and Huxley predicted and that compute things much as Hubel and Wiesel described. By arranging point neurons into hierarchies, taking care to exploit interactions between excitation and inhibition, as well as variations in the speed and strength of signals, model nervous systems appear able to generate complex patterns of activity like those in real brains and appear capable of computing basic facts about the external world.

This computer metaphor–guided approach has been extended to entire brain areas. In the visual system, a given region is seen as computing a particular quality based on responses earlier in the hierarchy. For example, areas where many neurons respond to visual motion are believed to compute this property based on spikes passed from V1. Some have even attempted to model consciousness itself as a computation performed in a hierarchy of spiking neurons.[20]

There is no question that the brain transforms signals to perform computations and that similar computations also tend to be grouped in the same brain regions. Such transformations are essential to brain function. In turn, almost all artificial intelligences from the 1950s to today succeed by adopting this basic framework. But brains do more than just compute: they also flexibly communicate. This fact has been obscured by the focus on single neurons and their computations. Our understanding of whole brains has been limited as a result.

Models of combinations of neurons are much less powerful than models of single neurons because our models of many neurons tend to view the brain as an idealized system built of point neurons. Most of the time, models are confined to a single domain, such as vision or memory, and typically just one subsystem such as color, motion, or place memory. When individual neurons are linked together in a network, the system as a whole takes on behavior that can't be predicted even with a full knowledge of each component. Instead, brain dynamics emerge in ways that depend on how the system is integrated, and on its global goals.

Just because we don't understand the emergent behavior of the brain, it does not mean we can dream up theories willy-nilly. It makes sense to start by building up simplified models of the brain from smaller units we do understand, such as single neurons. Moreover, we should certainly align our understanding with measurements from anatomy and physiology. But not every aspect of anatomy and physiology is necessarily helpful.

More fundamentally, it is not at all clear what the right thing is for anatomists and physiologists to measure. This is the purpose of theory: to define for the experimentalist the right things to measure. Theorists examine and remix the existing basis of knowledge into a new form. They determine which things matter and which don't. Then it is up to the experimentalist to figure out how to make the appropriate measurements. In neuroscience, we need a new theory to deal with the complexity of brains and to help experimentalists figure out what to measure.

A new theory of brains will take a long time, but there are clues about how to proceed. There are ways to approach the brain that differ fundamentally from the standard approach of computational reductionism. An alternative is to think holistically and to consider the constraints under which the whole brain must operate. This approach has a long history in neuroscience but has mostly taken a back seat to reductionist approaches.

Brains must work in a specific ecological context or habitat. Instead of asking how neurons can be put together to make brains, we can ask what the environment demands of the brain in terms of neurons. Here we invoke evolution, which shapes everything in biology to a greater or lesser extent. Assuming that evolution has found effective and robust solutions to the challenges of allowing an animal to survive and reproduce, we can ask what would be a good way to solve a particular problem. Then we can test whether real brains do something similar.

This approach is especially useful in sensory systems, whose job is to assess physical quantities and qualities of the natural environment. Brains must do this efficiently. Senses like vision consume disproportionate amounts of energy compared to other brain systems, and they are among the first to go if a brain is injured, say, if there is a lack of oxygen. Evolution has presumably found parsimonious ways to run a neural system for sensing the world. The key to efficiency is to know what to expect. The important question, it turns out, is: What is typical in terms of the pattern of light we receive from the world?

The idea of matching sensory systems to the patterns of photons that they are likely to encounter traces back at least to a 1961 paper by Horace Barlow, a neuroscientist who was also the great grandson of Charles Darwin.[21] Barlow, who died in 2020, was writing his seminal paper just as Hubel and Wiesel's landmark work was being performed. He focused on an earlier and better-understood stage of processing in the visual system: the retina. Like Hubel and Wiesel, he assumed that the basic solution for sensing light and processing it in the vertebrate retina is basically the same across species. The various neurons of the retina, we now know, grow and develop in the same order in all mammals and most vertebrates.[22] This includes photoreceptors, which sense light, and a class of neurons called ganglion cells that Barlow focused on. The retinas of all vertebrates possess similar types of ganglion cells, which relay visual information received at the eye to the thalamus (and then to the visual cortex). Any information not passed over ganglion cell axons will never lead to visual awareness.

Ganglion cells' job is to sample and evaluate a small chunk of the retinal image. This is the same concept—called a receptive field—that Hubel and Wiesel applied to simple and complex cells in V1, except that it is occurring earlier in the visual stream. The patterns of light that ganglion cells are looking for are also simpler than those of V1 cells; they measure donuts of light, so to speak, instead of rectangles (see figure 3.4).

Imagine looking down on a sugar-coated minidonut on a dark table. The contrast of a bright white donut on a dark background is the pattern that ganglion cells are looking for. At each point on the retina, there are lots of donut detectors of different sizes that overlap considerably, like a pile of donuts.

Patterns other than donuts but that resemble them somewhat—a pair of white stripes separated by a black gap and

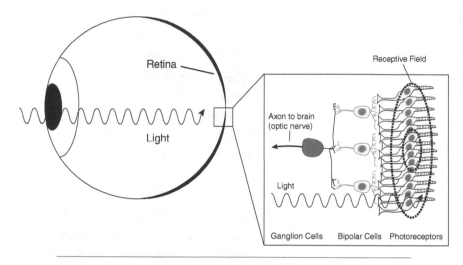

FIGURE 3.4 A ganglion cell of the retina has a receptive field shaped as concentric circles, shown here with dashed lines. This receptive field organization is called center-surround, with the center circle and the surround circle having opposite polarity (excitatory or inhibitory). These neurons receive signals from photoreceptors (via another class of neurons called bipolar cells). When a photon excites a photoreceptor, an electrical signal is generated that can contribute to activity in the ganglion cell. The ganglion cell summarizes activity across numerous photoreceptors. The ganglion cell's preferred pattern of stimulation (the pattern of light for which it produces the most spikes) is described by the receptive field. Axons of the ganglion cells form the optic nerve and communicate visual information to the brain's thalamus. Image by Daniel Graham.

surrounded by a black background—can also excite the ganglion cell, but to a lesser extent. The bright and dark parts of the scene just need to match up with the cell's desired pattern of stimulation. The donut-likeness of the pattern of light determines how many spikes the ganglion cell produces. Other ganglion cells look for donut holes: a white circle on a dark background. In reality, the donuts and donut holes these cells are looking for are generally much smaller than actual donuts and donut holes, especially in the central part of our visual field. But notice here that what the ganglion cells pass along to the brain is how donut-like a portion of the retinal image is, not the photoreceptor's direct recordings of photon levels.

A consequence of this coding strategy is that the brain receives information about relative light intensity across the visual scene, rather than absolute intensities at each point. A donut detector is really asking: to what extent is a central region different from a surrounding patch of light? This means discarding substantial amounts of photonic information.

At first blush, throwing away lots of detailed information from the photoreceptors in favor of a comparison measurement of light levels seems like a bad idea, especially so early in the visual system. Outside the central area of vision (called the fovea), each ganglion cell can subsume the responses of one hundred or more photoreceptors using this kind of pooling. The ganglion cell can't tell the difference between a point of light hitting one photoreceptor it is listening to and another photoreceptor it is listening to; all it knows is that it is excited by something in the world. But Barlow argued that this is a very good strategy—if we consider what the world typically looks like.

We tend to assume that the visual world is impossibly diverse, with myriad ecological and geographical arrangements. Everywhere we look is different, and wildly so. Here are some trees, there a river, and somewhere else an expanse of desert. Sometimes there are faces or elephants, sometimes there are flowers or snakes. In fact, practically any visual environment on the surface of the earth shares basic properties.

Barlow guessed that one of these properties is that neighboring points are similar to one another, and he turned out to be correct. We can test this for ourselves by pointing our finger in a quasi-random direction in our environment and assessing light intensity at the pointed-to location, as well as immediately to its right (or to the left, or above or below). On average, points chosen this way will look similar in terms of light intensity. In other words, it is a good bet—but not a guarantee—that neighboring points look similar.

This is precisely the wager made by the ganglion cells. Since similarity—correlation—is the norm, ganglion cells look for places where this *isn't* the case: bright areas next to much darker patches. When these areas wash across a ganglion cell matching them, the cells fire at high rates (around a hundred spikes per second). These patches turn out to be especially important in visual scenes because they tend to occur at the edges of objects.

So, as Barlow argued, encoding the visual world in this way— measuring only particular relationships among neighboring points, and throwing away much of the photonic information received by the eye—is a good way to start doing something useful with a visual image, such as finding the object-like stuff in it. It's good because it builds in knowledge about the world. The ganglion cells' solution is so good that no vertebrate can deviate very far from it and expect to see. In other words, evolution has demanded that all advanced visual systems deal with the peculiar visual structure of our environment using ganglion cells and their donut-shaped analyzers.

Claude Shannon's information theory predicts that an efficient code will remove predictable structure in the incoming signal. Since it's a good bet that the visual world has correlated neighboring points, an efficient strategy according to information theory is to reduce correlations when we encode this signal. This way, we can use fewer symbols to encode the same amount of information. To a first approximation, this is what the early visual system does. Barlow showed that information theory helps explain why vertebrate visual systems have ganglion cells that decorrelate.

That is, ganglion cells seem to encode the world with less correlation between neighbors than was present in the input.[23]

Notice that Barlow's insight did not require looking for new properties in neurons, or interacting with real cells at all. His theoretical understanding didn't even require any new data—only the linking of existing ideas. It nonetheless led to key advances in understanding how neurons behave in living animals.

The thrust of research Barlow initiated never gained a catchy name. It is referred to as efficient coding theory.[24] But Barlow's ideas give us a powerful example of holistic theory. The study of how evolutionary demands related to regularities in the sensory environment influence neural coding ended up providing many insights. Efficient coding theory is now in a sense gospel, forming a basic element in several contemporary theories of how brains work.[25] Efficient coding theory also suggested a crucial tweak that improved artificial neural networks and propelled them to the forefront of neuroscience research and AI, as we will see. Barlow's holistic approach can be a model for the next phase of neuroscience research on brain networks.

Until the 1990s, Barlow's ideas were still outside the mainstream in neuroscience, and he himself was equivocal about them. He performed electrophysiology experiments in mammals, as well as behavioral experiments on human perception, but didn't directly apply his insight about matching the brain to the world until other scientists had advanced his ideas much further.

Barlow's ideas were first tested when British neuroscientist Simon Laughlin applied them to the coding of light intensity by flies. Laughlin asked if neural computations of light intensity in the fly's eye were related to the distribution of intensities in the fly's environment. He showed that this was the case in 1981. Responses of cells in the fly's eye to light of different intensities were optimally matched, according to Shannon's information theory, to the distribution of light intensities in the fly's natural habitat.[26] In other words, these cells seemed to know what to expect in terms of the number of photons they would likely encounter from moment to moment.

To some extent, this is still neuron-doctrine, computer-metaphor thinking: Laughlin studied average properties for six cells, trying to deduce their strategy of computation. But using global analysis of operational environments as a theoretical tool was an important development. It was the first of many empirical studies to take this approach.

By the late 1980s, a few more scientists started to advance Barlow's ideas. In 1987 David Field in the United States and Geoffrey Burton and Ian Moorhead in Britain independently showed that neighboring points in our visual environment are indeed likely to be similar, just as Barlow had assumed.[27] In addition, these researchers found that this property did not depend on the scale of the visual environment: a small backyard and a wide vista have the same likelihood of neighboring points being similar.

The importance of Barlow's insight was increasingly recognized in some quarters in the early 1990s. Its significance grew with the discovery of what is called sparseness. It turns out that the world has other predictable structure besides similar neighboring points. The visual world is also sparse. Sparseness has a particular mathematical meaning in this context. The basic question is how to utilize a set of visual letters to describe the visual world we inhabit, including all the faces, elephants, mountains, deserts, and forests we can see. This is again a holistic approach, and one that subordinates the actions of single components to global, system-wide goals.

We can understand sparseness by analogy with how we write down a spoken language. Consider the code used in written English versus that used in written Chinese. English is represented with a few coding units—the twenty-six letters—that are combined in ways that require all or nearly all of the coding units in order to express any idea. Chinese, on the other hand, is represented using thousands of different coding units called characters. Each character is used relatively rarely but carries more specific information. In contrast, English letters carry much less information on their own, and knowing some of the letters in a word doesn't necessarily narrow down what the word is. The written English words *car* and *far* share two of three coding units and could be confused rather easily. But the Chinese characters for these words (车 and 远) would never be mistaken for one another.

English letters (derived from Latin letters) have the advantage of being able to encode most sound systems because the coding units correspond to individual speech sounds; transliteration is more cumbersome in Chinese because individual characters often encode several sounds (the word for *far*, 远, is pronounced *yuǎn*, consisting of three speech sounds in combination). But again, this comes at the cost of needing to use most or all of the coding units at least once to write out any idea in English.[28]

In contrast to written English, written Chinese is *sparse*: a given idea requires a highly specific set of characters. Over many ideas—a whole book, for example—Chinese texts will probably use most characters in the large "alphabet" or codebook at least once. But only a select few are needed at a time.[29] Sparse codes like Chinese and dense codes like English have advantages and disadvantages, and neither is inherently better as a code. The interesting question is: What kind of alphabet or codebook is used in the language of vision, and in the brain more generally? Is it more like English or Chinese?

The visual world, it turns out, is sparse. As a result, the brain's coding strategy for visual information—especially in the cortex—also appears to be sparse. Thus the neural code for vision is rather like written Chinese.

In a sparse code, each descriptor—a tilted rectangle of light at a particular location, like that of simple cells in the visual cortex— is used relatively rarely, but when it is used, its use is emphatic. In terms of activity in real neurons, we should see most neurons quiet most of the time—like the vast majority of Chinese characters—and only used or activated fully when needed. At any given time, we should see a few neurons highly active, rather than lots of neurons sort of active.[30]

This is precisely what neurophysiologists have found, and not just in the visual system. There is a similar sparseness in the auditory world, particularly in speech sounds, which is matched by a sparse encoding system in the auditory brain. In fact, almost any system of the mammal cortex, from spatial memory to touch to motor systems—seems to produce spikes in quite sparse fashion.[31]

Having a large library of quite specific coding units that fire sparsely turns out to be very advantageous in terms of energy use as well. Only 10 percent or less of the brain's neurons can be active at once. This is because making a spike is so costly in terms of metabolism. If the neural code were like English, we would be using almost all units at least once in practically every phrase. "Vowel"-like units would be in use almost continuously. If neurons encoded information in this way, many of them would need to produce at least a few spikes pretty much all the time. The body simply could not sustain that level of energy expenditure.

But just because the brain is sparse, it is not necessarily the case that sparseness is shaped by the environment. It took a clever experiment that exploited the sparse visual properties of the world to show that the brain does indeed exploit sparseness in the world to build its internal codes.

In 1996, David Field and Bruno Olshausen asked an artificial neural network to "grow" a codebook of detectors like those in the visual system. Each detector was a square of a few dozen greyscale pixels that started out randomly. Olshausen and Field fed patches of natural scene images to the network. The network, which learned in essentially the same way as the Perceptron, tried to reconstruct each scene patch using only a few detectors. Reconstruction means combining detectors in the codebook by superimposing them, scaling their contribution to the reconstruction, and multiplying them together pixel by pixel. This results in a new patch, which should come close to matching (reconstructing) part of the natural scene.

If too many detector patches were needed to reconstruct a given scene patch, the neural net was penalized by adjusting the pixel values in the detectors. In this way, the network adjusted the pattern of light each detector was looking for. Over time, the detectors got better at reconstructing the patches in a cooperative and sparse way. The detectors took on distinct shapes. They weren't donut detectors, as might have been expected. Instead, they were bar detectors just like the simple cells Hubel and Wiesel discovered. In fact, they were closer to the actual detector design of simple cells known from high-precision modern measurements (more like elongated ovals than the rectangles Hubel and Wiesel proposed).[32] See figure 3.5.

Olshausen and Field showed that two basic design goals, fidelity and sparseness, are sufficient to generate a visual code that matches the key properties of real neurons in our visual brains. It's pretty clear that neurons need fidelity—they should respond in relation to the actual pattern of light stimulation in the environment. But the importance of sparseness had not been recognized before the Olshausen and Field experiment.[33]

Their discovery would have wide-ranging implications. Within a few years, there was evidence from neurophysiologists

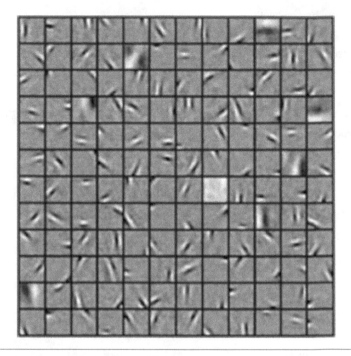

FIGURE 3.5 The array of receptive fields "grown" by Olshausen and Field using an artificial neural network architecture. These units or filters generate a sparse code of natural scenes: to reconstruct any given patch of a natural image, only a few of these filters need to be combined. Compared to these filters, real visual neurons in the primary visual cortex have similar preferences for spatial patterns; their combined activity also indicates that the brain uses a coding scheme whose activity is sparse.

Data from Bruno A. Olshausen and David J. Field, "Emergence of Simple-Cell Receptive Field Properties by Learning a Sparse Code for Natural Images," *Nature* 381, no. 6583 (1996): 607–609. Image courtesy of David Field.

William Vinje and Jack Gallant showing sparse activity in visual neurons of the primate brain when they experienced the natural visual world.[34] Prior to this, neurophysiologists almost never presented visual neurons with natural images, instead preferring the orderliness of rectangles and the convenient mathematical properties of related patterns called gratings (repeating bars of light and dark whose brightness varies in a sinusoidal way).

The finding that real neurons responded sparsely to the real world showed both that the brain is fundamentally sparse and also that sparseness is influenced by the sparse structure of the visual environment. This was an insight about brain components that came from studying the system holistically, rather than from a reductionist point of view.

One consequence of a sparse brain is that many neurons seem to fire very rarely, or perhaps never. This is a problem for experimentalists working in the computer metaphor framework, who are trying to understand what a given neuron is computing. If a neuron in the visual cortex fires rarely, we probably won't be able to find the precise kind of visual pattern it is waiting to see, and thus we won't know what its computational job is. It's a dirty secret in neural recording studies in live animals that many if not most cells researchers listen to with electrodes don't fire reliably in response to any external stimulus (be it photons to the eyes, loud sounds to the ears, bodily motion, etc.). Measurement or even the existence of these neurons was until recently routinely omitted in publications. But as far as anyone can tell, in parts of the brain dealing with vision, hearing, touch, motor, and cognition, very sparse activity is the norm.

The reality of a brain with lots of neurons with no obvious computational function that are often left out of research results has led to the notion of neural dark matter.[35] Dark neurons, like their astrophysical counterpart, are objects we know are there but whose identity and purpose we can only study indirectly. Sparse activity in the brain prompts deeper theoretical problems: if many neurons spike rarely, why build them at all? Neuroscientists haven't worked out a good answer to this question, but there is a discipline-wide effort to understand sparseness in neural systems.[36]

AI has also exploited sparseness. It turns out that deep learning is also fundamentally sparse. In particular, deep learning uses sparse codebooks to learn about inputs. Deep learning approaches—specifically convolutional neural networks (CNNs)—are largely concerned with growing a set of detectors appropriate for a given task such that the ensemble is very sparsely active. This is essentially the same goal as that of Olshausen and Field's sparse coding. But whereas Olshausen and Field's sparse coding applies to unrestricted natural images, deep learning applies the idea to a specific input set.

Instead of trying to encode any possible image, or the range of visual environments a sighted creature is likely to encounter, deep nets just try to encode those visual features that are relevant to the task they have been given, such as differentiating cars and airplanes. And they are forced to do it sparsely. This means that a deep net trained to distinguish cars and airplanes is going to learn a set of detectors that are active in differential fashion for images of cars versus images of airplanes. Each detector will be used rarely, but when it is used, it will resemble a diagnostic feature of the image set. For example, a detector might look like a somewhat distorted fragment of a license plate (something that would not appear in an image of an airplane). Responses across detectors will be sparse, like written Chinese. The responses of the detectors to other classes of images—trees, for example—will be ambiguous, since the net's detectors were not trained to distinguish trees from other things.

This strategy has allowed CNNs to far surpass earlier approaches in computer vision. Traditionally, the encoding system used in vision AIs has had a predefined code or alphabet. The alphabet in a computer vision system is a set of spatial patterns of dark and light. Some, like the SIFT algorithm, use donuts and donut holes of different sizes as their set of detectors, much like ganglion cells.

Other approaches have looked for lines of pixels that vary abruptly in space from dark to light, rather like simple and complex cells. Though these earlier approaches resemble—indeed, were often modeled on—the human visual system, none were designed around the need for sparseness in the ensemble of detector responses.

CNNs are different in this respect: they "grow" their own sparsely active detectors based on the task. Benchmark tasks like finding faces "in the wild" can now be performed at almost 100 percent accuracy by CNNs.[37] The same principles can be applied with similar accuracy to analyzing sounds such as human speech (with the time/frequency domain substituting for the 2D visual space of images).[38]

The success of CNNs in computer vision and related fields depends on other insights besides sparseness (including those of McCulloch and Pitts and Rosenblatt, of course, as well as insights from researchers like Kunihiko Fukushima, who saw the need for many layers of detectors, each tuned to increasingly complex image features; this is where the "depth" in deep nets comes from). But sparseness is the secret sauce. In turn, the success—albeit qualified—of CNN models in predicting the responses of visual neurons is also due in significant part to their assumption of a sparse coding framework.

Remember that we can trace these insights right back to Barlow's ideas, to measurements of holistic regularities of the system and to evolutionary arguments based on those measurements. The fine details of single neuron activity—the domain of the Hodgkin-Huxley model—are mostly ignored in this approach. Instead, the essential sparseness of neural activity—dictated by the structure of the world and the need to be efficient—has been found throughout the brain. No animals' brains were harmed in Olshausen and Field's experiment.[39] And yet we know much more about living brains as a result.

Still, much remains unknown about the brain. The neuron doctrine and its partner, the computer metaphor, have no doubt moved us forward. We shouldn't discard this theoretical framework. Instead, what we need in addition is holistic thinking. We have seen an example of the success of holistic thinking in relation to efficient coding. We can apply this kind of thinking in relation to the flexible communication systems in brain networks as well.

Next, we will examine how the connectomics movement is beginning to provide the infrastructure for an improved, holistic understanding of the brain. As we will see, the brain is, at a fundamental level, a massively interconnected network, and one ripe for investigation in light of the internet metaphor.

4

FROM CONNECTOMICS TO DYNOMICS

IT is fortuitous that neuroscience is currently engaged in a revolution in the holistic understanding of brains. This is the connectomics revolution. There is a rapidly increasing base of knowledge about large-scale network structure of brains from which to build new theories, especially those concerned with communication.

A key insight of connectomics is the concept of the connectome itself. Olaf Sporns, the neuroscientist who coined the term in 2005, has described the genesis of the idea, which came to him while preparing a review paper on brain networks.[1]

> The human brain is a complex network whose operation depends on how neurons are linked to each other. When attempting to understand the workings of a complex network, one must know how its elements are connected, and how these elements and connections cooperate to generate network function. The human connectome describes the complete set of all neural connections of the human brain. It thus constitutes a network map that is of fundamental importance for studies of brain dynamics and function.

Beyond foregrounding the brain's global interconnection, connectomics has pushed neuroscience forward largely due to insights centered on three interrelated themes: small worlds, hubs, and connection strengths. In turn, these concepts form key linkages for the internet metaphor of the brain.

A lot of effort in recent years has been devoted to mapping connections among neurons in different brain areas. The idea is to figure out what connections exist by tracing forward and backward from neurons in a given region.

To trace the connections of a neuron, we first need to make it stand out. A tracer is a molecule that highlights just a few cells rather than the whole forest of cells. We need to sneak a traceable chemical into the neuron without damaging the cell. This is often done by replacing a molecule in an essential protein with a radioactive version of the same molecule. After injecting these kinds of tracers in the living animal and letting them slosh around in the neuron, we remove the part of the brain we want to study. We then trace the radioactive molecules by adding fluorescent dye that glows in their presence. Images of the glowing axons are taken in successive thin slices of the brain tissue. The images are finally combined into a 3D map of the neuron and its environs.

In recent years, scientists have made substantial advances in tracing techniques, including with single molecules and also viruses. Genetic engineering of the rabies virus has allowed greatly improved tracing from dendrites back to incoming axons. Rabies is a particularly nasty virus because it only infects neurons, and does so very aggressively. These properties make it an excellent tracer. Advances have also been made in tracers that spread "backwards" from axons to dendrites, as well as in machines to cut up brains into thin slices in order to find neurons containing the tracer.[2]

One of the most obvious but underappreciated findings to emerge from tracing data is that we can get from anywhere to anywhere quite easily on the mammal brain network. This appears true at large and small scales: most brain regions are just a hop or two from most other regions. And a given neuron in a local chunk of brain is also just a few hops away from most other neurons in that chunk.[3] This means that any neuron in the brain is just a few hops from almost any other: comprehensive studies are lacking but the typical number of hops (synapses) between any two neurons is probably about 3 or 4.

As we saw in chapter 1, the architectural motif in mammal brains that creates these short paths between any two neurons is called a small world. It is characterized by relatively dense local connectivity, where local groups form clusters, and a relatively small number of long-range connections link members of each cluster. Scientists debate the exact definition of the term *small world*, and they also debate whether the brain meets any of these definitions. But all evidence supports the conclusion that short paths of just a few synapses separate any neuron and region from almost any other, in significant part because of the pattern of local and long-distance connections. The closeness of every neuron to every other has been suspected since at least the mid-1990s[4] but has now been confirmed using tracers.

What's equally clear is that there is no central control: there are no switchboards that exert control over the whole system. In bigger brains like ours, the network relies instead on hubs. In a small world, loose local clusters link with each other over a few longer connections. But on top of this structure, there are a relatively small number of nodes that interconnect many other nodes. These hubs also tend to be connected to each other. In other words, hubs connect to other hubs.

This organization has been called a rich club: like an oligarchy, those with the most bountiful connections readily share those connections with each other, cementing their power.[5] A lack of comprehensive data and disagreement about mathematical definitions have made it hard to discern exactly how much human brains form a rich club. In smaller brains, such as that of the mouse, hubs can be almost fully connected to one another. With fewer neurons in the cranium, and consequently fewer brain regions, it is more feasible to connect all hub regions to each other.[6] The mouse brain appears to take advantage of this fact. But full interconnection of hubs is less feasible in much larger human brains. In any case, interconnection of hubs (to a greater or lesser extent) helps put every neuron just a few synapses from any other neuron.

Connectomics is now in the midst of a new phase of investigation to map more connections among brain areas with very high precision. A key figure in these efforts is neurobiologist David Van Essen of Washington University in St. Louis.[7] In the early 1990s, he and his postdoctoral student, Daniel Felleman, assembled a map of connectivity in the macaque monkey visual system from past reports in the anatomical literature.[8] They found reliable reports of 305 connections among 32 cortical areas involved in vision. The iconic diagram of these connections has become known as the Van Essen Diagram (figure 4.1). Its spaghetti of connections has glazed the eyes of many undergraduates over the years. But the knowledge contained in the diagram has taken on a new urgency in the connectomics era. Previously, without the notion of a connectome—and without knowing much about how the visual network fit into the wider network of the brain—it was hard to make much sense of the diagram.

At the bottom of the Van Essen Diagram (which I have left unlabeled to give a sense of its structure) are shown the inputs to the visual system from retinal ganglion cells, via the thalamus. The topmost area is the hippocampus, whose connectivity to the visual system was not studied by Felleman and Van Essen. The top of the diagram is not meant to indicate the "top" of the brain, or

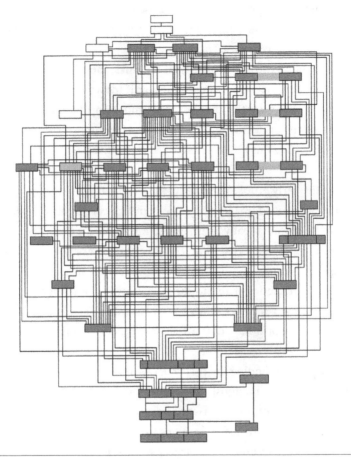

FIGURE 4.1 The "Van Essen Diagram" showing connectivity among
subregions of the brain that compose the primate visual system.
All connections shown are one-way (up in this diagram), though many
return connections (downward) also exist. Signals travel among the
regions, but in ways that are not yet well understood.

Adapted from Daniel J. Felleman and David C. Van Essen, "Distributed
Hierarchical Processing in the Primate Cerebral Cortex," *Cerebral Cortex* 1,
no. 1 (1991): 30.

any kind of end point. Indeed, neuroscientists have long debated whether the vertical alignment of the various areas in the diagram makes any sense. And although Felleman and Van Essen's data included the directionality of the links, their iconic diagram leaves out connections going "backward" (downward in the diagram; all links in the diagram transmit information upward). Inconsistent protocols in the underlying anatomical studies that Felleman and Van Essen collated led to further confusion. What was needed was a comprehensive study.

In the early 2010s, Van Essen, in collaboration with an international team led by Henry Kennedy in France, launched a major effort to systematically map the connections suggested in Van Essen's eponymous diagram. The goal was to assemble the full range of connections in a single experiment with a consistent protocol. In a series of ambitious axon-tracing studies in mice and monkeys, the team of Kennedy, Van Essen, and many colleagues around the world has substantially expanded our understanding of the mammal connectome.

One major finding in this work is that core regions of the brain are indeed almost fully interconnected. This discovery builds on the notion of network hubs and rich clubs, which had primarily been detected in humans using less precise methods of brain imaging in humans. In tracer studies, far more precision is achievable. Studies of the mouse performed by Kennedy and colleagues showed that hub regions include key areas that handle touch, auditory, and motor systems. Each of these regions can reach every other area over a single hop. The twelve core areas Kennedy and colleagues studied had connections to all eleven other areas (save for a single missing connection).[9]

This organization seems like a good a way to get information from any area to any other over a minimum of hops. Thus, each large grouping of neurons (touch, motor, olfaction, and so on) can talk to every other one directly. And it means that most of the mouse brain's other two hundred–odd areas need not be connected nearly as densely as the twelve hubs.

In the mouse, the hub regions are related to the animal's lifeways. Mice explore the world using tactile sensation in their whiskers and also are highly attuned to auditory and smell information; their visual systems are relatively weak because they are nocturnal and live in dark places. However, mice will resort to vision in daylight if they are starved. The relative utility of sensory modalities appears to be reflected in which areas serve as hubs. In the mouse, only one primary sensory area for vision was fully connected to the other core regions, while touch, auditory, and motor systems included more than one fully connected hub.[10]

Connecting every core region to every other core region becomes less tenable in bigger brains like ours and those of other primates. As brains get bigger, the number of neurons grows, of course, but the number of distinct areas also increases, and hence many more connections are needed to interconnect all of them. Full interconnectivity of core regions seems possible only when the number of core regions is around a dozen. With seventeen areas defined as hubs in the monkey brain, Kennedy and colleagues found fifteen missing links that prevented full interconnectivity of the core. But despite lacking full interconnectivity in the core, monkeys and other primates still have very short paths between any pair of nodes. In addition to hubs, Kennedy and colleagues also confirmed that regions are very likely to be connected to their neighbors, and that a small but substantial number of longer axons link local clusters.[11]

Thus, Kennedy and colleagues have eliminated any doubt that the mammal brain—including that of primates—is designed fundamentally around the goal of getting messages from anywhere to anywhere else very easily. This is achieved not by connecting every neuron to every other (as in successive layers of deep net) but by linking hubs with most other hubs; by connecting local groups of regions to each other; and by employing shortcuts. Moreover, it is just as clear that many different types of information interact directly with one another, and therefore must be interoperable.

Kennedy and colleagues, as well as other groups, have also started to measure how strongly one area is connected to another. Beyond asking what is connected to what, the question is now, what is the anatomical *weight* or strength of the connection in terms of the number of links between two nodes? Investigations of how many connections exist have spanned small scales (measured in terms of the number of synapses between two neurons) and large scales (measured in terms of the number of axon fibers between two areas). Measuring weights is possible because of massive efforts using tracers in rodents and monkeys.[12]

Weight data prompt a host of new questions about how brain parts exchange information. As it turns out, the brain has a special form of connectivity in terms of weight that emphasizes neighborhoods. When two populations of neurons are nearby in space, they are not only more likely to be connected to each other, but they are also likely to have stronger (weightier) connections.

At small scales, such as local networks in the visual system, most connections are made by a relatively small number of individual synapses (in the low hundreds). A minority of linkages between neurons involve lots of synapses (thousands). Researchers have also mapped the intricate connections among neurons in different cortical layers, and of different neuron types, with similar results. These studies, though significant, come from fewer than forty neurons labeled with tracers, and they involve significant assumptions about unobserved connectivity patterns.[13] But it does seem to be the case that there are typically a small number of strong connections between nearby neurons, and a lot of relatively weaker connections with somewhat more distant ones.

At larger scales, we know more. The goal is to trace long axons, which can be a few centimeters or more in length, to see how larger areas are interconnected. Ideally, we would find a few synapses in a given area and trace the axons involved back to their area of origin elsewhere in the brain. If we could squirt the same amount of tracer into each area and see how many axons it highlights, we could get a relatively unbiased measure of how many axons connect each little chunk to each other chunk. The problem is that tracer chemicals don't stay exactly where we put them. Even a small squirt diffuses across anatomic boundaries. So at present we can only get a sense of relative connectivity, unless we make a lot of assumptions about brain architecture (as is done for local neuron-to-neuron weight estimation). But measurements of relative weights—given by the fraction of all neurons labeled in an experiment that connect a given area to another— have revealed important insights about the connectome. This is a major advance, which has been in the works for decades.

With their massive anatomical data, collected from thousands of neurons in dozens of monkey and mouse brains, Kennedy, Van Essen, and colleagues were able to construct a simple and quite accurate model of the relationship between how far away any two areas are and how densely they are connected. Statistically, they found a distribution of connection weights that falls sharply with distance but has rather broad shoulders.[14]

Take, for example, the pattern of connections bringing information into the primary visual area, V_1, of the monkey, located in the back of the head. The question is how many axons come into V_1 from an array of other cortical areas (organized in terms of how far away they are from V_1 in 3D space). Leave aside for a moment the fact that classical notions of V_1, predicated on the computer metaphor, largely ignore these kinds of "backward" connections. Indeed, the computational picture (and the iconic Van Essen Diagram) sees V1's job as primarily about sending information, rather than receiving it.

The data of Kennedy and colleagues show that an area directly adjacent to V1—V2—accounts for a very high proportion of all cortical axons going into V1 (figure 4.2).[15] Areas just a bit further

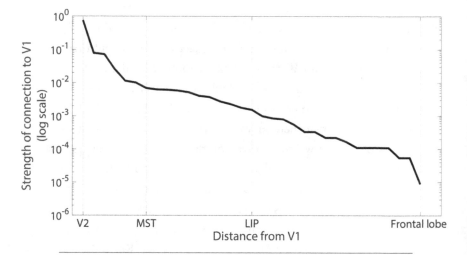

FIGURE 4.2 Relative strength of connections from different parts of the brain to V1 (primary visual cortex). Areas like V2 that are near to V1 in space have strong connections with large numbers of axons synapsing in V1. Area MST is only a bit further away but sends far smaller (though still substantial) number of axons to V1. An area considerably further away, like LIP, sends fewer axons still, but the number is of the same order of magnitude as for MST. The furthest areas, like area 8 of the frontal lobe, send 100 times fewer axons than LIP. However, though these long-distance axons are few in number, they appear to be crucial for spreading signals widely and quickly in the brain. Despite the cost and difficulty of constructing region-to-region connectivity in this complex way, the relationship of connection strength and distance appears to hold for all mammal brain areas. MST: medial superior temporal area; LIP: lateral intraparietal area. Image by Daniel Graham.

Data from Nikola T. Markov et al., "A Weighted and Directed Interareal Connectivity Matrix for Macaque Cerebral Cortex," *Cerebral Cortex* 24, no. 1 (2014): 17–36.

away, such as area MST (the medial superior temporal area, which is classically seen as a computer of visual motion), send less than a hundredth as many axons to V1 (the y-axis in figure 4.2 is logarithmic: each major hash mark is a power of ten larger than the one below). In other words, moving only a little bit beyond the local neighborhood of V1 reduces the number of axonal connections drastically.

But there is a broad shoulder to this distribution. In figure 4.2, we can see that the curve stays fairly flat from left to right for a long time after the initial dropoff and before the final fall. Beyond the local neighborhood, many areas each contribute a smaller but significant number of axons at a variety of distances. Areas like LIP (the lateral intraparietal area) are a few centimeters further north of V1 compared to MST. Though LIP is further away from V1 than MST is, the strength of the connection from LIP to V1 is still more than 20 percent of the strength of the connection from MST to V1.

The most distant areas measured, such as a part of the frontal lobe of the cortex (if we were wearing a baseball hat, this is about where the logo would be), are quite a long way from V1, which is at the back of the head. The frontal lobe sends only a tiny number of axons to V1: the strength of the connection is only around 1 percent of the connection from MST to V1, and less than 0.001 percent as strong as the connection from V2.

These long-range connections are especially costly and difficult to build in large brains like ours. Longer connections also mean significant delays in signal travel time. But these costs are worth it because brains gain an ability to share information widely—and, as I will argue, flexibly.

The overall pattern of a broad-shouldered distribution of weights holds for essentially all the brain areas Kennedy and colleagues traced. This work has shown that mammal brains adhere to a basic commitment to send information over long distances—even as this gets harder and harder to achieve in big brains.

Researchers at the well-funded Allen Institute in Seattle have investigated similar questions using similar methods and have largely come to the same conclusions. Like Kennedy and Van Essen's team, the Allen Institute researchers have been able to measure the strength of connections among brain areas using tracers. Focusing on exhaustive studies of the mouse brain, Seung Wook Oh, Julie Harris, Hongkui Zeng, and numerous coworkers have found that most areas of the cortex are only a single hop away from one another.[16]

Another contribution of the Allen Institute researchers has been to map the full set of connections between areas of the cortex and subregions of the thalamus (which are called thalamic nuclei). They have shown that the thalamus is a critical hub, one that can be thought of as part of the brain's network backbone. The thalamus sends and receives a great variety of messages between most of the major cortical systems of the brain. Thus, while hubs in the cortex connect to other hubs in the cortex, they also connect to the thalamus and back.

The connectivity of the thalamus is worth considering in detail. It allows us to imagine the ease with which messages can traverse the brain, since the paths all go through a common structure. Signals are relatively canalized in most nuclei of the thalamus: axons from a given cortical area group together when they arrive in a particular nucleus (though they do send side-branches called collaterals to other nuclei along the way).[17] For example, the nucleus subserving information from the eyes is called the lateral geniculate nucleus; it only handles visual information. But in connecting regions, there is substantial spreading of signals among systems. In the cortex, as we have seen, hub areas are well connected to each other, meaning that visual information interacts directly with, for example, motor information. Because of this interaction, signals can easily get to other parts of the cortex

by way of the thalamus. In addition, following trips to the thalamus, signals can return to a different part of the thalamus and
travel on to still other parts of the cortex.

Based on the data from Oh and colleagues, we can trace a path
through axon bundles of the mouse brain to see how easy it is to
get around the various realms of the brain. (see figure 4.3). Starting arbitrarily in the auditory cortex, we can get to an area of the
thalamus called the lateral posterior nucleus, and from there we
can travel back to the cortex to any of several areas devoted to

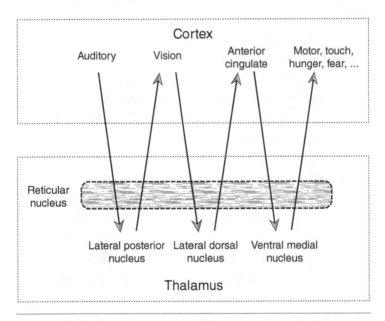

FIGURE 4.3 Connectivity of areas of the cortex to the thalamus.
The top of the diagram shows inputs from cortical areas to various nuclei
of the thalamus, while the bottom of the diagram shows outputs from the
thalamus to the cortex. Each arrow represents a bundle of axons.
The dotted rectangle labeled RT indicates the reticular nucleus of the
thalamus, a structure of neurons through which cortical inputs and
outputs pass. See text for descriptions of the functions of cortical areas
and their thalamic targets. Image by Daniel Graham.

vision. Departing vision, and passing back through the thalamus (lateral dorsal nucleus this time), we can make for the anterior cingulate cortex, an area not well understood in the mouse, but believed to be partly involved in generating motor outputs.[18] Now we could continue our journey from the anterior cingulate cortex via the ventral medial nucleus of the thalamus on to other cortical motor-related areas, or to cortical areas involved in feelings of hunger, in fear responses, or touch sensations. Each time, we travel by way of the thalamus. Almost every trip to the thalamus can be a turned into a return trip through reciprocal connections. Return routes on paths of more than one hop are also viable and may be shorter than forward routes.

It's important to note that the thalamic nuclei we are describing here are relatively canalized in the sense that incoming messages from different sensory systems are largely kept separate. The paths we are following involve lots of parallel wires.[19] However, each of these wires connecting the cortex and thalamus passes through the reticular nucleus of the thalamus, which can modify signal traffic by sharing information among parallel wires. The densely interconnected reticular nucleus of the thalamus is like a bun holding two plump hot dogs (the hot dogs, one in each hemisphere, contain the other nuclei of the thalamus, including lateral posterior, lateral dorsal, and ventral medial nuclei; see figure 4.4). Reticular neurons provide dense interconnections among neurons traveling to practically all nuclei of the thalamus. Consequently, virtually every neuron in the thalamus that exchanges messages with the cortex is influenced by thalamic reticular neurons. As such, they could be important for flexible network control and routing. In addition, the reticular nucleus could deliver acknowledgments (acks) of message receipt, an idea we will consider in more detail in chapter 6.

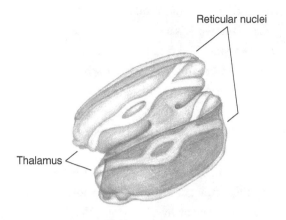

Reticular nuclei

Thalamus

FIGURE 4.4 Anatomy of the thalamus's "double hot-dog bun" shape. The main masses of neurons constituting the thalamus look a bit like a pair of short, plump hot dogs that are joined at their midsections. The reticular nucleus is a neural structure that cradles the hot dogs on opposite sides. Axons traveling to and from the thalamus pass through the bun-shaped reticular nucleus and interact with neurons there. Image by Reanna Lavine.

Most of the connectivity between the thalamus and the cortex in mice was known before the work of the Allen Institute scientists. But by mapping the entire connectome of the animal in a single experiment, they were able to place the thalamus in the context of a wider array of brain networks. Studying the entire network also allowed measurement of connection strengths between areas, as with the work of Kennedy and colleagues. Oh and colleagues have confirmed the finding regarding the broad-shouldered distribution of connection strengths between brain areas as a function of distance.

Going forward, weight data like those from the Allen Institute researchers and Kennedy's team will be increasingly valuable to connectome researchers studying network structure, and will surely be useful in many other interesting ways (in chapter 8 we will discuss recent dynamic modeling work that exploits these connectomes). This work is also a model for large-scale collaboration to study a system holistically.

Progress in the field of connectomics has also been achieved by researchers like Henry Markram and the Blue Brain project, which is building extremely detailed computer models of small chunks of the cortex in the mouse. Others, such as the Human Connectome Project, funded by the National Institutes of Health, are using brain imaging in an attempt to map human brain connectivity (tracer injections being obviously off-limits for mapping connectomes of live humans). We will consider these efforts in later chapters.

Olaf Sporns's concept of the connectome remains a work in progress. Despite great progress, we currently lack comprehensive tracing data at small scales, so our picture of the network as a whole remains limited. Part of the problem is that local structure is very complicated: different cell types are interdigitated and arranged into an extraordinarily complex weave of layers. Revealing the connections among cells in a small neighborhood thus requires labeling different cells with different markers. It's also important to label only a small fraction of cells; otherwise the traces overlap too much to be disentangled.

We are a very long way from mapping the networks of even a small fraction of the neurons in the mammal brain, and even further from understanding their functional significance. This is why it is too soon to ascribe functional significance—say, distinguishing between healthy and disordered states like schizophrenia in humans—based solely on current measures of network architecture.[20] Indeed, it's not enough to know how brain networks are structured. We also need to understand the organization of dynamic activity across the whole connectome. This emerging area of research is sometimes called *dynomics*.[21]

We can start to make progress in understanding the dynome if we ask what kinds of communications are best suited to the brain network architecture we are learning about—and to the environment in which brains operate. From an increasing basis of knowledge in connectomics, I am advocating a holistic approach of asking what a network of neurons must do to deal with particular communication challenges, as engineered by evolution. The approach complements the reductionist approach of asking what behavior is possible by assembling well-understood single neurons into a complex grand architecture. In turn, I believe the insights we glean from holistic studies of brain networks can help us understand single neurons even more fully. Call it Reticular Theory 2.0.

. . .

We need to understand the constraints on brain networks and the communication challenges they face. Today we know that network characteristics are not fully apparent in single neurons and the computations they perform. The simple fact is that one cannot tell what message-passing protocol is in use by looking at the rate of spiking activity in a single neuron in response to some stimuli. What appear to be spikes related to computations on sensory inputs might, when considered part of whole-brain interaction, be more related to communication goals. For example, the signal may be related to where a subsequent message is going, or it may indicate that a previous message was received. As such, a set of spikes doesn't necessarily correspond to content (such as "glowing rectangle, up and to the right"). These kinds of signals may be misunderstood if viewed solely from the point of view of the neuron doctrine and computer metaphor.

We are lucky that a system we ourselves engineered—the internet—has "evolved" to solve similar holistic challenges as the brain. Moreover, just like the brain, the internet's success depends critically on the essential unity of the network, where any node is just a few hops from any other and can pass many different types of information.

We can study computer networks like the internet independently of the internal workings of individual computers. What matters is the signals that are passed: how many, when, where, and how. Single computers are obviously separate entities, just like single neurons. But connecting them together on the internet produces something altogether different, and interesting in its own right. However, each quirk of the computer is not necessarily relevant to such holistic behavior.

As we will see, the basic message-passing strategies of the brain and the internet have common constraints. For one thing, they are both sparse in their activity, but they share other characteristics as well, such as the need for reliability, scaling, and, perhaps more than anything else, flexibility. But to understand the internet's special relevance to brain network communication, first we need to look closely at the internet's infrastructure and why it is designed as it is.

5

HOW THE INTERNET WORKS

WITH the advent of Web 2.0, attitudes toward the internet have evolved and changed. Machine learning and AI continue to play an increasing role. With all the clamor over covert influence campaigns, data breaches, and the like, utopian visions of an internet-enabled future are no longer tenable. Yet the internet is playing a more and more important role in all of our lives, for better or for worse, especially since the start of the coronavirus outbreak.

But in all the tumult over what we do on the internet, it can be easy to overlook how revolutionary and remarkable the internet's basic message-passing technology is. Network researchers Romauldo Pastor-Satorras and Alessandro Vespignani summarize the internet's goals this way:

> The internet is not driven by any supervising agent or authority, nor follows the blueprint of a pre-established architecture. It grows and develops because of cooperation and self-organization, to conform to technical standards and associative needs. Indeed, if we look at the internet on a coarse grained scale, we see a spontaneously growing system, whose large-scale dynamics and structure are a cooperative effect due to many interacting units aimed at optimizing local communication efficiency.[1]

This chapter presents a nontechnical introduction to the workings of the internet and of communication systems more generally. The same basic technology underlies both the internet proper and cell phones. The aim here is to describe the "physics" of the network—the overarching principles that form its conceptual superstructure. We won't dwell on specific technological mechanisms, which in any case change fairly rapidly over time. As we will see, it's the internet's general principles that make the system so powerful, and they are also relevant to the brain.

To the extent that we think about the infrastructure of the internet at all, we might imagine server farms, satellites, and fiber-optic cables. These elements are important for the internet to function optimally but are not essential. Although electricity is its currency, and computing power makes it hum, the most important technology of the internet is the scheme for making sure that messages are delivered. The most truly novel and critical technology that powers the internet is its unique set of *rules*.

The rules that govern the internet are known as *protocol*. They are decentralized and public. The rulebook largely consists of documents known by the initialism RFC, which stands for *request for comment*. The name for these documents is itself a testament to their open, collaborative origin, since their principal purpose from the 1970s to today has been to solicit feedback for the formulation of revised, better rules. In retrospect, it is hard to imagine how a self-organized global communication network like the internet could be built in the first place except by input and consensus from users. One cannot point to any other world-spanning technology that was built this way. Because it directly takes account of input from users, the internet is not only efficient at transmitting information widely but is also multipurpose. As more users joined, they found that the open architecture allowed for a great variety of uses. The internet remains a uniquely successful creation of the hive mind. As an adaptive system, it may even display features of consciousness, an idea we will return to in chapter 9.

The public rules that must be followed to join the internet are straightforward, flexible, and cheap to implement in hardware. The great innovation of the internet's rules becomes more visible when the internet is compared to other communication systems that allow dispersed parties to pass messages to each other.

Communication systems must be able to exchange information widely and flexibly. Each system operates on a network, or an arrangement of nodes and links. Nodes can only interact with other nodes—such as by sending a message—if they are somehow connected to one another. The wider system gives each node the opportunity to communicate at will with any other node via some set of links.

Whether we're talking about the *yam* postal system of the Mongol Empire eight hundred years ago, or the French Minitel computer network of the 1980s and 1990s, all communication systems require protocol. The Mongol postal system, which spanned much of Eurasia for centuries, only worked because each station adhered to the protocol of keeping horses ready for messengers at all times and knowing how to get to the next station. Routing protocol is no less important in modern communication systems. At its height in the early 1990s, the French Minitel system reached millions of users before being superseded by the internet, which uses a similar protocol. Minitel protocol required that information be contained in packets of fixed size.

In any two-way communication system that humans engineer, the global rules largely determine success or failure. And while communication protocol often involves computations—such as deciding where to direct a signal using logical operations like IF and OR—it does not require computers per se.

Routing is fundamental to a system. Maybe a better term is *queueing*: the real task is dealing with multiple signals and putting them in line. Indeed, if there were only one message at a time, routing would be unnecessary. Engineers like Leonard Kleinrock of UCLA, a pioneer of the internet, have likened routing to a system for lining people up to get on an airplane. Boarding happens via a channel, namely, the cabin door. A channel in a communication system can be a conduit, wire, or cable, or a particular electromagnetic frequency band—anything that can put signals in a line. The channel of the cabin door has limited capacity: it has to be shared by many "signals" or, in this case, people. Sharing is not always easy. As Kleinrock put it in his landmark 1976 text *Queueing Systems*: "Recently, I made the mistake of flying across the country in a Boeing 747. As a queueing systems analyst, I should have known better!"[2]

Why not just make the channel bigger? Boarding a plane is often faster when more than one cabin door is used. This is called *multiplexing*: having more than one channel running in parallel, which allows more signals to pass. We might think that, in a perfect world, routing could be obviated by multiplexing. Why worry about how messages are directed and sent across the network when we can just make really big doors or tubes? There is definitely something to be said for multiplexing. Our mail carrier's satchel, which carries many letters, is just as much a multiplex unit as a plane's cabin doors or a fiber-optic cable, which carries messages on many wavelengths of light. But the ability for any communicator to reach any other communicator on a large network requires nodes capable of sending messages on more than one output route. The only way around this challenge is to have every node make a direct connection to every other node. This solution is untenable for any large-scale communication network, just as it is for the brain (recall that our brains would be 20 kilometers in diameter if they had all-to-all connectivity like this).

Lacking all-to-all connectivity, messages will inevitably collide. If we do nothing to redress collisions, colliding messages are likely to be lost. This is obviously a problem if we want to deliver messages reliably. We need to have signals line up in a queue so we can figure out where they each need to go and then send them there. Having bigger tubes doesn't solve this problem. But a good set of queueing or routing rules can.

All routing protocols involve trade-offs that affect how people can use the system and how communicators are connected. There are surely trade-offs of communication engineering in brains, too, which are reflected in rules for how messages are lined up on the network. In the brain, these rules probably don't look like the rules of airplane boarding—nor do they look exactly like those on the internet. But some rules must exist for routing in the brain, and the principles behind these rules are, I argue, employed for similar reasons as on the internet.

Starting in the 1960s, electrical engineers suggested that there are three basic forms of routing protocol. Each is best suited to particular regimes of network connectivity, information density, and timing. A *circuit-switching* protocol is a system that allows for dedicated links between every pair of nodes. The name refers to the aim of establishing a circuit, like a metal wire that can pass flows of electrons in either direction (depending on which end of the wire they are "pushed" from). A computer's internal communications involve literal circuit switching, with circuits called *buses*. As the name implies, a bus carries any collection of passengers on a specified route. A bus connects memory chips to central processing units (CPUs), for example, or a graphics card to a display. Buses can deliver large tranches of messages at close to the speed of light, but only between two nodes at a time. The circuit-switching protocol of a bus is a simple and effective way of moving ones and zeroes over a predetermined route using electrical current.

Circuit switching doesn't require computers, or even electricity, however. Human-to-human vocal communication can also be considered a kind of circuit switching. Protocol is about the rules, not the physical substrate. Circuit switching requires an exclusive, two-way channel of exchange. It also requires the ability to "call" or secure the line of communication between parties. The communication protocol for a band of *Homo sapiens* hunter-gatherers could be imagined as circuit switching. Most of the time, a few dozen members of a band can communicate back and forth with any other, since everyone is usually nearby. To call someone, we just need to shout their name. Any time one person is talking, others usually refrain from talking at the same time. In some ways, circuit switching remains the state that all human communication systems aspire to.

Traditional telephony, as practiced through the mid-twentieth century, is the classic engineered form of circuit-switching protocol. It was important because it introduced intermediaries in the channel called switchboards. Every phone had a dedicated wire to the switchboard, and each switchboard had wires connecting it to other switchboards. Since signals are electrical and can flow in either direction, only a single wire was needed to connect each phone to the switchboard, while many wires running in parallel connected switchboards to each other. Signals could be sent at will from either end of the wire by simply generating current on the wire. The wires connecting two parties were seized for the entire period of a call, excluding all other users. It's a system whose network architecture and routing protocol grew out of the way people used the technology, which was to converse intimately.

Above all, telephony assumes *synchrony*. Both communicators must be in sync, or coordinated in time. This worked because people want to speak directly to one another in real time and to share a tiny laugh or sigh, talk over one another, or leave silences. The telephone remains popular for those seeking sex or business negotiations or mental health assistance.

With telephony, we can reach directly into the life of any potential communicator as long as we know their address on the network (today it's a phone number, but in the past simply knowing the person's name and city was sufficient). Once we dial a person's number, they have a brief window to open the channel by picking up the phone. Then the channel can be maintained indefinitely. But to achieve this level of reach and intimacy via circuits, the system sacrifices some efficiency and flexibility. With traditional telephony, when two people are on a call, no one else can reach either of them directly. And whether they are speaking or not, no other signals can be sent on the wires connecting two people—nor could they send signals other than sound.

Another way to organize two-way communication systems is called *message switching*, and it is typified by postal systems. The communication channel in this case is a single human mail carrier who is responsible for all addresses in a geographic area. In this scheme, the message's integrity and reliability are paramount and take precedence over intimacy and speed. When we send a letter, we wrap it in a protective envelope and hand it off to a trusted channel: the mailbox. We may even ask to receive confirmation of the message's travels across the network. In past eras, message integrity was ensured by unique, tamper-evident seals. Once it arrives, we can be sure that our message has been received in exactly the same form it was sent.

But to achieve this level of integrity and reliability, we sacrifice intimacy. Love letters are nice, but they are usually a surrogate for the real thing, which is best carried out in person—or at least over the phone. Message-switching systems like the post also demote speed because mail is usually picked up and delivered only once a day. Postal mail is an *asynchronous* system: communicators are not passing messages at the same time. This follows from the network's architecture.

Communication engineers today recognize that the problem of routing has to do with how much we share communication channels. But this fact remained mostly obscure until the computer age. For most of human history, channels of communication have been unitary: they passed a single stream of information, to the exclusion of all other messages, and they were sent all at once. If I am talking to you on the phone, I can't carry on a simultaneous conversation with someone else. Postal systems allow somewhat more sharing of the communication channel: my mailbox and the mail carrier's satchel can deliver several communications simultaneously. But postal systems retain the unity of the message. If I mail you a long letter, there will be no reason to put each page in a separate envelope.

Once we delegated the task of person-to-person communication to computers, however, new approaches became possible. Messages could be subdivided in any arbitrary way. We are still exploring this new universe of routing protocol. Indeed, it is not clear how big the space of possible protocols is. Humans have not necessarily dreamed up all of the basic arrangements yet. The brain may use a routing protocol that is utterly unlike those we have engineered. But the essential success of the internet's revolutionary routing protocol—called *packet switching*—is self-evident, and for this reason it deserves careful consideration in relation to the brain. By understanding the power and flexibility of the internet's protocol, we can gain a novel perspective on how the brain intercommunicates.

Several innovations in computing and networking led to the early development and later success of packet switching. The first innovation was time sharing, or the ability to share computer time with more than one user. In the mid-twentieth century, the only computers were mainframes, which required laborious procedures to load and extract information. By the early 1960s, computer engineers had streamlined this process, in part through the invention of programming languages, which allowed more than one user to perform computations. But a key shift was purely conceptual: instead of computers being a simple daily service operation like the neighborhood dry cleaner, they could be organized to perform computations for many users, and to do so twenty-four hours a day. This change, due largely to the American psychologist-turned-computer-scientist J. C. R. Licklider, also allowed users to queue their computer tasks so they could be scheduled for a future time, when the user need not be present.

Mainframes remained solitary, however, until the most sophisticated computing operation—that of the U.S. military—started to have several mainframes in the same building. In 1966, Bob Taylor, another former psychologist, was hired to run the Department of Defense's Information Processing Techniques Office at the Advanced Research Projects Administration (ARPA). Taylor's office included one of the most advanced computing facilities in the world. He looked at the three mainframes he oversaw and wondered whether they might be able to talk to one another. Each computer had a separate terminal, but neither the terminals nor the mainframes they controlled could communicate with each other. At the time, universities and other organizations were also clamoring for computing power. Given the enormous expense of building and maintaining a mainframe, there was, if nothing else, a national security interest in making sure they were utilized efficiently.

This practical problem required two theoretical advances, which both came from Paul Baran in the United States (and were discovered in parallel by physicist Donald Davies in Britain) between 1960 and 1965. Baran, then a researcher at the RAND Corporation in California, had a background in electrical engineering but was working at the time in the realm of pure communication theory. The problem he was working on was ensuring survivability in nuclear missile command. With such dangerous weapons deployed over large areas—both in the United States and the USSR—the danger was that individual missile posts might lose contact with central command during an attack and not know whether they should launch. Robust communication was essential.

Baran's first insight involved how the command network was wired up. The way a network is connected—its *topology*—shapes the ways it can be used. In a diagram that evokes Herbert Spencer's of a century before (figure 2.8), Baran saw network organization as spanning a spectrum from star-like connectivity to fully local (lattice-like) clustering as shown in figure 5.1. As seen on the left, a network could have every node connected to a central node. But relying so heavily on a central node would make the system very vulnerable. It wouldn't be robust to attack. At the other end of the spectrum, on the right side of the figure, connecting nodes only to their closest neighbors as in a lattice makes it hard to get information from one place to another quickly. An important message would have to pass through many nodes to get from one end of the network to the other. Each hop introduces potential errors and failures.

Baran realized that the key was to compromise between these extremes. If nodes had a few long-distance connections *and* collected signals from a local neighborhood, as in the center of the figure, the network could achieve speed as well as robustness. This pattern achieves the same kind of connectivity as what would later be called small world–like networks.

Baran calculated that with only three or four times more than the minimum number of links needed to interconnect the

Centralized (star)   Hybrid (small-world)   Localized (lattice)

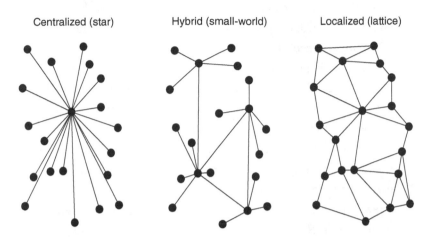

FIGURE 5.1 In communications networks, connectivity can span a range of organizational schemes using the same array of nodes. At one extreme, the star-like structure on the left, all nodes are connected to a central node. In this case, any message must pass through the central node to reach its destination. At the other extreme, on the right, each node can only communicate with its nearest neighbors. Paul Baran envisioned a third type of organization, shown in the middle, that compromises between these two extremes: nodes are connected to their neighbors but also possess a few longer-range connections. This requires more wires than the star network but greatly reduces the number of hops a message must take compared to a nearest neighbor network. Image by Daniel Graham.

system, we can achieve very high robustness to node destruction. At this point—1964—Baran made a crucial decision. Rather than shielding his insight behind a patent or in a classified archive, he chose to publish it in the open scientific press. His rationale was humane but also practical. As he said, "Not only would the U.S. be safer with a survivable command and control system; the U.S. would be even safer if the USSR had a survivable command and control system as well!"[3] Baran set the tone for the evolution of the internet by making it open and collaborative.

In theory, any viable routing protocol can be used on just about any network topology, but topology constrains the kinds of routing protocol that will work well. As we will see in relation to brains, it is not enough to know only the topology of brain networks in order to know how the network operates. But brain network topology (along with the functions brains must perform) can be used to suggest effective and efficient routing schemes.

This is where Baran's second insight comes in. He needed a way to pass messages over the small world–like distributed network. The system he envisaged was one where each node was only a few hops away from any other node but there were no central switchboards. Routing reliably on this kind of network required a leap of faith. He had to trust the whole network, rather than a switchboard. Further, he had faith in the benefits of randomness, as we will see. Baran's solution, developed with computer scientist Sharla Boehm,[4] came to be called *packet switching*. The name was provided by Davies, who independently devised essentially the same scheme in 1965, about a year after Baran and Boehm's work.[5]

Packet switching involves chopping messages into small pieces and releasing them separately onto the network. As shown schematically in figure 5.2, a message such as an email is divided into equal-sized units and labeled with which part of the original message each unit contains. An application like an email program simply dumps a set amount of data from the message into the requisite number of packets.

It's not immediately obvious why we would want to do this. The reason is that once messages have been packeted, we have a lot more flexibility. Parts of different messages can be mixed together. Rather than devoting every phone line or envelope to sending a single, whole message, we can now potentially include lots of different messages along the same channel, squeezing out a lot more information-carrying capacity in the channel. It's

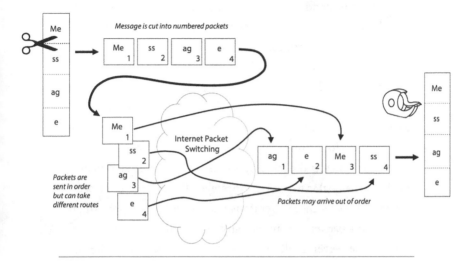

FIGURE 5.2 Schematic description of packet-switched communication strategies. A message is cut into equal-sized pieces, each labeled with its position in the original message. They are sent across the network, possibly on different paths (though typically they take the same path). However, they may arrive out of order. The receiving machine reconstructs the original message. Image by Daniel Graham. Adapted from James Gillies M., and R. Cailliau. *How the Web was Born: The Story of the World Wide Web* (New York: Oxford University Press, 2000), 22.

multiplexing in time, and it is an extremely efficient solution. Packet switching works best if our channel is digital and the time domain can be divided into very small chunks.[6] This scheme also requires asynchrony.

The next question is, how should the packets be lined up for transmission? When messages are chopped into packets, each is tagged with the address of the sender and receiver. As with postal mail, the address is inspected at every node along the way. But postal systems tend to send messages on fixed routes. Each post office has a specified path for all messages destined for a given target. If I were to bring a letter to the post office, it would not make sense to hand it off to whichever mail carrier happened to be walking out the door at that moment. But the first approach to queueing Baran proposed did something like this. He called it the "hot potato" method of directing packets across the network. Each node would have a *routing table* giving directions from that node to others on the network. When a message comes into a node, the node tries to send the message on by the fastest route. If any links along this route are busy—even momentarily—the node experiencing the blockage tries other routes as quickly as possible. The goal is to pass the message off like a hot potato, even if it might take a more circuitous path to its destination. Importantly, each packet can potentially take a different route under this scheme—even if all of the packets are going from the same sender to the same receiver.

Packet switching welcomes randomness and indeterminacy, and it was a major departure from all communication systems that had come before. To make this scheme work in any practical sense, packet switching required small computers at each node. Even with hot potatoes, we need to store messages in a queue, however briefly, and to figure out how to get them to their destination successfully. This necessitated the invention of routers, or minicomputers whose only job is to handle the flow of messages on their network neighborhood. Like so much else in the design of the internet, the idea of a router (due largely to computer scientist Wesley Clark) seems obvious today. But it came in an era entirely devoid of PCs, or really any electronics more complex than a transistor radio and less complex than a mainframe.

Within five years, Baran's solution was recognized and exploited by Bob Taylor's ARPA team and put to practical use as the first communication network for computers, called ARPANET. Taylor's team worked with Davies and Kleinrock on the system's design while the Stanford Research Institute (SRI) planned the physical linkages via copper wire and the Boston firm Bolt Beranek and Newman built the routers.

On October 29, 1969, Kleinrock's lab sent the first packet-switching communication between his lab in Los Angeles and SRI in Palo Alto. The goal was to "log on" to the machine in Palo Alto in the sense of being able to give standardized commands to the Palo Alto computer from Los Angeles. After the first two letters of the log-on command "LO" were received in Palo Alto, the system crashed, but it was fixed and fully operational later the same day. And Lo, there was the internet.

Though the first application of the internet was to connect two computers in order to share computational resources, packet switching is very useful in other contexts, too. Today, it underlies both the internet proper and ubiquitous cellular networks. But packet switching is not strictly dependent on computers or other hardware. Rather, it is a flexible, robust, and efficient way to communicate among many parties across large areas. When we focus on communication goals and protocol, as early packet-switching engineers did, we may find effective strategies that would be illogical or invisible if we only worried about computation.

Copper wires are by no means required for packet switching. Following ARPANET, a second packet-switching system was built as a radio network. ALOHANET, in Hawaii in the early 1970s, connected university campuses across the islands. Radio networks have limitations because of their broadcast nature. Unlike a wire, a radio signal radiates over a wide area and is not private. ALOHANET had one central station and several receiver stations. Since everyone within range of the central broadcast station received the same set of signals, some kind of order needed to be imposed. For ALOHANET, this meant a strict schedule for the central broadcast: at regular intervals—and only during a brief window—signals would be sent from the central station that were intended for a particular receiver.

But traffic flow in the opposite direction—from receiver stations to the central station—was possible anytime, thanks to another key innovation that had to do with avoiding collisions of signals. When two peripheral stations tried to send messages to the central station at the same time, both messages would be destroyed. The rule about how to proceed next was that both peripheral stations would have to wait a random amount of time to try again. This is called a *back-off algorithm*, and it worked splendidly. After colliding, both signals had to back off before trying again. Because the back-off time was random, signals were very unlikely to bump into one another more than once.

The back-off rule was applied uniformly to all packets. It had the effect of supercharging the routing system. ALOHANET's back-off algorithm took the indeterminism of Baran's hot potato method and put it center stage. Not only could individual routers be trusted to pass lots of individually labeled message chunks, but when there was a collision, the network's architecture could be trusted to inject random noise into the system locally to become more robust. Waiting a random amount of time to resend messages after a collision remains the critical trick used on local network protocols such as Ethernet.[7]

I am highlighting the benefits of random noise here because most models of neural systems include a noise factor. In brains, noise is assumed to be caused by interference of electrical signals, imprecision of chemical mechanisms, quantum fluctuations, and other influences. What if noise in the brain were functional—a way to make the whole system communicate better?

We know now that random variation can improve the performance of many physical systems, for example, in what is termed *stochastic resonance*.[8] But in terms of communication, this idea is antithetical to Shannon's theory, which assumes that the goal of any communication channel is to minimize the effects of noise. As we will see in the next chapter, noise is simply too widespread a phenomenon in brains not to have some functional role. Perhaps it serves to facilitate intrabrain communication, as it does on the internet, as a kind of back-off algorithm.

Like ALOHANET, modern Wi-Fi is another form of packet radio communication, and it uses a similar approach to routing. Electromagnetic signals traveling between a device's wireless modem and the Wi-Fi base station are broadcast across a local area, but since electromagnetic signals add together causing interference, we have to make sure they stay in their lanes, so to speak. So Wi-Fi divides up the electromagnetic frequency spectrum into reserved bands for each device in the area. Every device connected on Wi-Fi sends and receives signals on a private frequency. But at the base station, there is a queue of messages. Signals reaching the base station are subjected to back-off algorithms much like that of ALOHANET. If there are collisions between packets from different devices, senders need to wait a random period of time before trying again. Modern Wi-Fi adds another trick. Before any packets from a given device are accepted by the base station, the station waits and listens to see if any other devices are also trying to send packets. A device on another Wi-Fi channel that tries to send packets will have to try again later, and each time the system waits and listens. This trick ensures that at least one stream of packets will be accepted by the base station and transmitted on, rather than blocking all messages in a collision.

The wait-and-listen approach is based on—but in practice more effective than—ALOHANET's randomized back-off method. But because Wi-Fi frequency channels tend to be seized for a long time—whether or not they are sending information—this leads to the same problem we saw in relation to old-fashioned telephony. Once claimed, Wi-Fi radio channels become unavailable to others wanting to join. Yet they may be passing no information. We are back to the problem we saw in relation to circuit switching: idle communicators wasting bandwidth. Currently, the U.S. DARPA agency (successor of ARPA) is investing heavily in solving precisely this problem in radio communication.[9] The goal is to allocate chunks of the frequency spectrum for Wi-Fi (and cellular networks) based on need rather than on the mere presence of a device. The nickname of this technology is *cognitive radio* because the idea is to allocate information channels intelligently.

Like the internet itself, clever engineering solutions to cognitive radio might someday be useful reference points for the brain since the brain utilizes some radio-like communications, where signals can be broadcast to a larger community of receivers. Indeed, we might learn about the brain from metaphorical linkage to many large-network communication strategies. But to do so we first need to see the brain's job as communication, not simply computation. The key is to ask: what are its trade-offs in network-wide communication, and what are efficient solutions to this problem? Understanding the internet's bag of tricks is a good place to start.

•   •   •

Even with effective strategies for dealing with message collisions, all communication systems need a way of dealing with disruption and errors. All kinds of events can stop a message from being delivered over communication channels, including power outage, router or switchboard malfunction, severe weather, squirrels, and careless or malicious operation of heavy equipment. Rather than attempting to prevent every possible disruption, most systems adopt some form of delivery verification and message resending.

From ARPANET to today, the internet has used a simple but highly effective strategy called the *acknowledgment* system, or *acks* for short. After a sending machine puts a tranche of packets on the network, it waits for an ack—a small return message—from the next router or routers. As far as the network is concerned, the ack is just another packeted message. Acks have to get in line like all other packets. When the ack arrives back at the sending machine, this confirms that the packets were delivered. If the sending machine waits a little while—125 milliseconds on ARPANET, a few microseconds today—and doesn't receive the expected ack, it will send the tranche of packets again. Some internet applications like the video chat protocol WebRTC also use *nacks* or negative acknowledgments (also called *naks*), which serve to alert the sending machine that expected packets were *not* received.[10]

The acknowledgment approach works because the internet is fundamentally *asynchronous* like the postal system, but also fast. The internet, despite decentralized organization and the need for constant back-and-forth acknowledgments, operates at extremely short time intervals. Routers have clock speeds fast enough to divide each second into a billion units. The specialization for fast, asynchronous communication also leads to highly sparse activity over the network, as we will see. Like the brain, most parts of the internet are idle most of the time, but when a message needs to get through, it leads to a brief flurry of activity. Modems, routers, and channels all tend to be active in bursts, operating near full capacity for a short time, then going mostly silent. This should sound familiar: brain networks are likewise highly sparse. We will explore this linkage in the next chapter.

• • •

Packeting a message into small chunks is absurd in an analog world. It would be like calling someone repeatedly and speaking one word of a message, then hanging up and doing it again. But in a fast, digital world, packeting opens up vast new territory in network communication strategies. It allows the system to easily and quickly check for errors, and it gracefully remedies collisions. Packeting also allows messages to find a good route based on current network traffic, as we will see.

We wouldn't have discovered packet switching without computers. Likewise, we probably won't understand intrabrain communication without understanding neural computation. But the design principles of communication in the brain will remain obscure unless we acknowledge that they don't necessarily follow from the computer metaphor. Internet-like communication requires ideas that are alien to the design principles of single computers.

Packeting allows interoperability. A packet can contain any digital data, as long as it is wrapped in a standardized envelope. Though ARPANET's design did not have the capacity to run applications other than passing simple text messages, the proto-internet was revamped to allow many possible applications in 1974, just five years after being turned on. Initially, the different applications that were harmonized were simply variant schemes for handling packets. But by the 1980s interoperability of packets was increasingly exploited as a way to do different things with packets. Communicators could then pass email, request and download remote files, and eventually surf the web and have real-time video chat.

The power of packet-switching routing protocol is that it allows both asynchronous and synchronous activity. An email, for example, exploits asynchrony to guarantee that all packets eventually arrive at the destination. We sacrifice some speed because the sender has to wait for acks from the receiver, but we can be virtually assured that the message will arrive in exactly the same form it was sent, just as in postal systems. On the other hand, synchronous communications like video chat can use the same packet-based system if they are willing to sacrifice some fidelity. Video chat exploits the internet's speed to deliver real-time communication at the cost of some lost information. Some packets are allowed to be lost during a video chat, but most will be delivered in less than the blink of an eye. Our senses are not acute enough to notice most of these lost packets. Though imperfect, video chat can simulate synchronous face-to-face communication and give a semblance of that old intimate connection humans crave.

The internet also allows broadcasts, or the delivery of signals from one source to many receivers, such as a streamed video. This works because individual parts of the stream fan out to a distributed network of memory allocations (caches). Everyone on the network is pretty close to some set of caches, but each part of the video follows a more or less unique path to a particular viewer. Making sure that every part of a video is available in a way that allows the parts to be put back together in real time is an ongoing architectural challenge, so here the trade-off is between fidelity, speed, and the complexity of the broadcast architecture.

It's harder to distribute signals in this way for live video, since there can only be one possible ordering of packets (now, followed by whatever comes next). This leads to yet another trade-off, in this case between fidelity and congestion. But the bottom line is that the packet-based system is flexible. Users can trade off among priorities such as speed, fidelity, complexity, and congestion as needed.

• • •

In a technical sense, the internet we know today is just one layer of a *stack* of routing protocol. As the name implies, the protocol stack is made of layers, and each layer interacts only with the layers immediately above and below. Packets move up and down the layers of the protocol stack in each machine along their route.

Let's say we have just pushed *send* on an email to our friend in another city. The email sits atop the stack at this point, in what is called the application layer. The application in this case refers to the kind of message, which could be a webpage (http), a file transfer request (ftp), or in this case, an email. As soon as it leaves our email program, the email is chopped into dozens of packets. Along with its siblings, packet #39 is first passed to the transport layer as it leaves our computer. This is possible because another application called border gateway protocol has been in constant contact with the transport layer through a parallel channel.

In the transport layer, packet #39 intermingles with packets belonging to other computers in our organization or network. This is also the layer that generates and keeps track of acks. From here, packet #39 descends to the internet layer of the stack, which is the internet proper. The internet layer's main job is to plan out a route for packet #39 and all the others (we will discuss the clever method for finding routes later in this chapter).

The contents of packet #39 may be rewrapped at this point to allow interoperability among networks. But the packet will retain the same basic information. The internet layer does a lot of the heavy lifting as far as message delivery is concerned. From here, packet #39 needs to be translated into electrical signals, which constitute the bottom of the stack. Though it happens very quickly, any message we want to send has to be loaded digit by digit onto the transmission line as ones and zeros. Here at the physical layer, there is protocol for handling electrical voltage pulses signifying "one" or "zero."

How does a router at the other end of a link know that the signals it receives constitute packets? The answer is that each router in the system inspects batches of signals for signs of packeting. For example, to identify a sequence of ones and zeros as belonging to packet #39, the system may look for a special code of 128 "ones." This marker indicates to the router that a prescribed set of information will follow, including the destination address, message size, timestamp, and payload. If the packet is in order, it will be treated equivalently by all internet routers between the sender and receiver.

Even after several hops to additional routers, the contents of packet #39 will arrive in exactly the same form as when the packet was sent. Each router has a ministack to accomplish this. Since the router doesn't need to read out our email, it will only pass the electrical signals from packet #39 up to the internet layer so they can be directed on the right path, then descend back to the physical layer for transmission.

When packet #39 arrives at the destination, it ascends the stack just like it would at an intermediary router. But instead of being directed back out across the network, the address on the packet tells the internet layer that the packet is at the right destination and should go up to the transport layer. From here, packet #39 is inspected for a signifier indicating the application that will use it, and the transport layer delivers it to that application. Other packets arriving around the same time, like those for a media stream, are sent to their respective applications, too.

Now back at the application layer, packet #39 rejoins its siblings, which may have arrived out of order, and the whole email is reconstructed by combining packet payloads together in their original order, which is easy since the numbering travels with the packet payload. Transmission of interoperable signals is thus a matter of descending and ascending a hierarchical stack (figure 5.3).

• • •

Network Topology

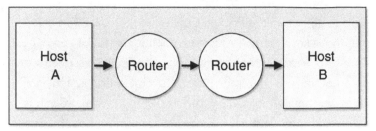

Data Flow

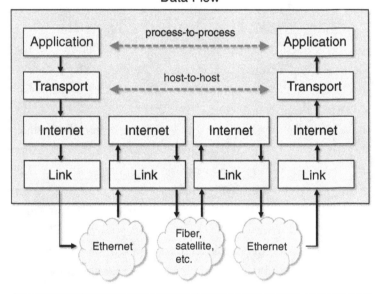

FIGURE 5.3 The basic design of the internet protocol stack. In terms of network topology, a message travels from one host to another via a network of nodes. However, the flow of data across the network is organized into conceptual layers. A message originates in the application layer and descends by way of the transport and internet layers to a physical link (wire, fiber-optic cable, electromagnetic signal). At intermediary routers, messages ascend only to the internet layer for route planning, then return to the physical layer for further travel. Image by Daniel Graham. Adapted from https://en.wikipedia.org/wiki /Internet_protocol_suite , CC-BY-SA 3.0.

The stack gives us the grand architecture of the internet. It describes the way messages move in and out of a given computer. The rules apply globally. Efficient operation of the internet requires a variety of other engineering tricks that also must be applied uniformly across the network. These tricks need to be fairly simple and easy to implement, but also flexible. We saw how the internet deals with the problem of collisions, which was by injecting randomness into the system. It ensures message delivery with acks. And it deals with the need for interoperability using the stack. The problem of pathfinding, which we will deal with now, demands a different set of tricks.

One trick is that the system exploits the small-world properties of the network in order to find good routes. Every local neighborhood is just a few steps away from any other. But more than this, each local neighborhood is given leeway in optimizing local message passing, as long as certain global rules are followed.

From a given machine (phone, computer, printer, etc.), the first physical connection is from the machine's modem to a router, which might be a Wi-Fi router in the hallway outside our office. These routers are all typically connected to a central hub router, often one operated by an internet service provider such as a telecommunication company (telco). One or more hub routers will have links to most or all communicators under their purview. In this context, there is usually no need to work out the best path for packets to travel within a telco's network since they can go through the hub and arrive anywhere in one hop.

The hub router is the boundary between a collection of machines in a subnetwork and the wider internet. Routers for more than one telco talk to one another via exchange points where wholesale exchange of packets can occur. Neighboring telcos may also have links between routers inside their respective networks. These private shortcuts—called "peering" links—can reduce transit time for packets because they don't need to pass through big exchange points.

If we want to get packets beyond our immediate subnetwork, our local router will need to know how to get there. This is a critical part of routing protocol that has to do with finding a good long-distance path. To find paths, routers talk among themselves. Every few seconds, they send tiny *keep-alive* messages to all the other routers that they are directly connected to. These small automated messages—a cousin of the ack—serve to let neighboring routers know that the sender is in service and working properly. Each router keeps a running list of neighbors that were in contact recently. If keep-alive messages arrive too slowly, receiving routers will assume that the sender is faulty and will not include it in potential paths. A router will pass keep-alive messages in a constant rhythm to let the neighbors know that it is ready to send and receive packets. It's like the heartbeat of the internet. A rather more useful biological analogy for our purposes, which we will explore in the next chapter, is that keep-alive messages are like the occasional spontaneous activity in the brain, which keeps the whole brain network in contact.

All routers must also respond in a timely way when queried. For example, if a router wants to know if another router at some distance away is functioning properly, it can send an *echo request*, a small message to which the receiving router must respond. Sending keep-alive signals and responding to echo requests are such crucial mechanisms that no router may join the internet without performing them. The information gleaned lets each router know which other routers are nearby on the network, and how to get there. Each router also then knows which other routers have the fastest connections, since the transit time of the request is also relayed back to the sender.

From this information, it is easy to work out several good paths between our local router and just about any possible destination. Lists of paths to popular destinations, called *routing tables*, are built from keep-alives and stored on each router (table 5.1). Up-to-date routing tables are also passed to all of the router's neighbors at regular intervals.

TABLE 5.1 Route of packets from the author's home router in upstate New York to the host google.com.

Hops	Name (IP Address)	Round trip time (milliseconds)
1	dsldevice.home (192.168.1.1)	1.794 ms
2	152.43.81.2 (152.43.81.2)	9.064 ms
3	201-9-77-30.static.firstlight.net (201.9.77.30)	8.427 ms
4	86-191-22-136.tvc-ip.com (86.191.22.136)	16.277 ms
5	be21.nycmnyqobr1.ip.firstlight.net (77.104.52.8)	18.759 ms
6	core1-0-2-0.lga.net.google.com (206.188.55.12)	44.822 ms
7	111.87.223.18 (111.87.223.18)	102.357 ms
8	172.253.70.18 (172.253.70.18)	60.536 ms
9	google.com (172.217.13.140)	33.524 ms

Hops 3, 4, and 5 occur at exchange points where many subnetworks are interconnected. Entry number 5, for example, is an exchange point in Albany, New York. Internet protocol (IP) addresses have been altered for privacy. Time spent in transit to a destination and back (roundtrip time) is listed in milliseconds in the rightmost column.

Routing tables include flags noting which routes have recently changed, making it even easier for this timely information to be propagated. It's a bit like regularly syncing our phone's contact database with the databases of all of our contacts. It is another example of the open architecture of the internet. To join the internet, we only need to follow protocol, such as sending keep-alives and responding to echo requests. This information is sufficient to figure out efficient routes. Routing tables that collate this information are value-added maps of the network neighborhood. They are provided for us at no cost by our neighbors. Each neighbor is a gateway to the wider network, running on the same open architecture. Once we join the network, the internet's full power is delivered to us for free, all the time, in the form of up-to-date routes. This kind of sharing benefits everyone. A similarly flexible and global protocol would be of great use in brains as well: each neuron's little computations could be integrated with those of many others.

Routing tables allow the internet to use the best route available at the moment. The particular routing scheme used by many routers is called Open Shortest Path First, and it is just what it sounds like. A router counts the number of hops required to get to a destination. It knows how many hops are involved thanks to its up-to-date routing tables. The router simply chooses the shortest one that is open or available at the moment.

What if the destination is many hops away from our local neighborhood on the network? In this case, routers may not know a full route to a particular target machine, especially if the destination is in another country. But the internet has a solution for this, too. It has to do with the design of addresses—internet protocol (IP) addresses.

Consisting of standardized groups of four numbers each, IP addresses are hierarchical, much like addresses in postal or telephone systems. The first few digits of our zip code, or the country code and area code of our phone number, greatly narrow down our location on the respective networks. IP addresses consist of four groups of numbers delimited by periods, with each number ranging between 0 and 255. The first numbers of an IP address identify the domain of the host, usually a large organization or telco. If we are in the United States and the destination of a packet is Japan, it is only necessary to find an exchange point somewhere near the location listed in the first numbers of the IP address (there are unlikely to be peering shortcuts in this case). Once the packet gets to a router at the exchange point, the chances are good that the router there knows a short route to the destination, since it continuously probes its own local neighborhood. Each local area of the network is well connected, and each cluster has some links to other clusters. Thus, the journey across the network is largely a matter of getting from cluster to cluster, or neighborhood to neighborhood, just as Paul Baran envisioned.

• • •

The design choices of internet engineers have shown a clear desire to plan for long-term survival. The internet was dreamed up to literally survive nuclear war. Computers, on the other hand, aren't like this. Most are used for less than a decade and are then replaced. They are discrete, disposable entities. By design, the internet was everywhere, making it robust.

The internet's distributed nature makes fundamentally altering its basic operation very hard. Luckily, its inventors had tremendous foresight in making its protocol both public and highly adaptable. Global robustness is a result of efforts at the local neighborhood level to pass packets efficiently, under global protocol that is open to all comers. Because of this openness, the internet has so far proven to be highly robust.

The internet has grown so quickly that it is rapidly outgrowing the four-number IP address, which is limited to around 4.5 billion unique values.[11] With billions of intercommunicating elements, the internet is already as big a network as the cortex of many monkey species and in the not-so-distant future might approach that of our own cortex, or even our whole brain. Its protocol was built to last and to grow. Even with increasing numbers of malicious actors on the network, and with the surge of activity during the coronavirus pandemic, the internet survives and even improves.

Brains, too, need to operate over long lifetimes—and even longer evolutionary trajectories—without fundamental changes to basic protocol or network structure. Thus, a first lesson of the internet metaphor for the brain is that a vast and growing networked system can succeed in the task of flexible message passing when two conditions are met: global rules must be followed, while the implementation of those rules is a local affair. We will look in more detail at specific insights of the internet metaphor in the next chapter. As we will see, these strategies may help the brain achieve robustness and growth.

The designers of internet protocol could not have foreseen every possible challenge of today's digital communication environment. These days, our internet-dependent culture is concerned with questions such as how many messages are sent by humans as opposed to bots. And even though we designed the internet, we don't really know basic facts about it. How big can it get? What is the actual routing behavior that messages experience? How long does it take for a typical message on a given path to reach its destination? Computer scientists and physicists have attempted to study these questions but with limited success. The problem, as with the brain, is in part how to take a representative sample. For one thing, probing the whole network or large parts of it is forbidden—even in the name of science—since it typically involves illegal botnets. But daily experience tells us that the internet just works.[12]

This success is largely due to the internet's most basic protocol, which, unlike physical law, is flexible. Likewise, we have no universal governing equations in biology or neuroscience. There are no laws that hold always and everywhere for biological systems. But there is basic protocol in biology and neuroscience, just as there is on the internet. And routing protocol, whatever it is, operates in brains such that—like the internet—it just works. Now we need to figure out what that protocol is.

As I have emphasized, the internet is different from the brain in many ways. The internet has a common clock, which runs at a very high speed; the brain, on the other hand, has patterns that recur over time but no fast-ticking, central clock. In addition, transmission costs are extremely low on the internet today, generally less than 1 percent of the total energy cost, since sending photons down a fiber-optic filament to relay a message uses very little energy compared to passing a spike down an axon to deliver a message.[13] But the internet's basic tricks for routing messages remain highly relevant for the brain.

The next chapter weaves together these tricks and their potential implementations or analogs in the brain. Beyond the basic organization of the internet described already, we will look in more detail at how the hierarchy of signaling systems that link neurons together mirrors the internet's hierarchy of message-passing rules. We will also look at how the internet's system for managing signals could be a good solution for a brain whose energy budget is severely limited. Coming up with viable theories of how the whole brain works may require creativity and a return to first principles—but in relation to network-wide communication, rather than just to computation.

6

THE INTERNET METAPHOR

First Steps to a New Theory of the Brain

WE have seen that brain metaphors are almost inevitable and that the currently dominant metaphor—the computer—may not help us in understanding flexible communication in the brain. The internet, on the other hand, has attractive parallels to the brain. How exactly can we implement and take advantage of the internet metaphor?[1]

This chapter is organized around four questions. These questions are not normally asked in relation to the brain under the computer metaphor but are especially relevant when we think about the network-wide communication requirements of the brain:

- How does the brain deal with collisions between messages?
- How is reliable delivery of messages achieved?
- How is flexible routing accomplished?
- How do we add more message-passing nodes?

Before we answer the first question, we need to think about where to look for evidence of the parallels between the mammal brain and the internet. There is a rich array of unexplained or ignored phenomena in neurobiology that may be better understood from an internet metaphor perspective. To understand why, we need to further examine the assumptions of the computer metaphor.

The dominant theories in neuroscience today are not just about computing; they are about *optimal* computing. As we saw in chapter 3, deep learning artificial neural networks can become arbitrarily good at representing a set of data. With enough computer power and time, a deep net can become optimal in a task like differentiating cars and airplanes in a particular set of photos. Neuroscience guided by the computer metaphor typically assumes that the brain is similarly optimized.

In a way, this is evidence that efficient coding theory has become too successful. Neuroscientists have seen that sensory systems and indeed the whole brain are shaped by the environment they operate in, and they have taken this idea to its natural conclusion. With universal computational machinery like that of artificial neural networks, the thinking goes, the brain can match the system to the environment as well as theoretically possible.

That brains are optimal is reflected in theoretical frameworks of the day. One popular framework is called the Free-Energy Principle.[2] Though couched in complicated mathematics, it really boils down to the same idea as efficient coding theory, just taken to the limit. The idea is that the brain takes note of regularities in the inputs it receives from the environment—patterns of color, motion, sound, social cues, and so on—and tries to build models of those regularities using neurons. The Free-Energy Principle says that the brain seeks a model that predicts new inputs in optimal fashion. It tries to make ever better predictions of the future on the way to optimality. Even consciousness is encompassed by the framework, explained as the brain's model of its own internal states. To use the jargon of the framework's proponents, the brain seeks optimality by minimizing the "free energy" in the system, which is essentially the difference between prediction and reality.[3]

While there is little question that brain systems are efficient, they are not necessarily performing optimal computations. In fact, optimality in most domains seems unlikely. Evolutionary biologists have been dealing with this question for a long time— really since Darwin proposed his theory of evolution. They have a much longer experience evaluating the notion of optimality. One view that has emerged over many decades of debate is that, in practice, an animal species cannot become optimal at something because over evolutionary time, the environment changes in ways that can't be predicted or managed based on past experience. Forests become grassland, glaciers advance or retreat, and populations must move, change, or perish. Strategies that are too specialized make it difficult for a species to adopt a different strategy when its ecology changes. The upshot is that species don't usually achieve optimality in a given trait because they will die off before it can be achieved.[4] Most are generalists, at least part of the time. Even species like the Galápagos finches Darwin studied, which have beaks that are highly specialized for their particular ecology, can adopt generalist strategies at times, and they are just as equipped to do so as less specialized finch species.[5]

In evolutionary biology, optimality is also hard to define. What constitutes an optimal eye? Researchers have long searched for tasks for which the eye is optimally effective. For example, they have considered the question of why our eyes (and those of other primates) have the specific sensitivity they do for red, green, and blue light. The textbook theory is that our eyes' color sensitivities are optimal for finding ripe fruit (typically reddish) hidden among foliage (typically green). But the biological and behavioral evidence shows that primate species in many different ecological niches all possess very similar color sensitivities. Moreover, primate eyes do much more than find fruit: they have to balance daytime vision and nighttime vision, as well as provide sensitivity to motion, all while supporting the ability to shift attention, guide movement, and identify predators, among other things.[6] When we start to consider higher cognitive systems in the brain as optimal solutions, the problem of isolating a single optimizable ability becomes intractable.

Thus, optimality isn't usually achieved in practice because every system has many interlocking components that together do many jobs. And even if optimality of most traits could be achieved, it probably wouldn't help the creature survive and reproduce for very long. Today, many evolutionary biologists (but by no means all) believe that nature might sometimes achieve optimality in certain systems, but evolution more often settles for solutions that are "good enough," and many if not most traits are selectively neutral.[7]

Since system-wide optimality of each trait is untenable, one of the main places we can look for evidence for an internet metaphor framework is in the messy stuff that is ignored in the hypothetical quest for optimality in neural computation. It is in the leftovers, the interstices, the blind spots of current theory that we may find evidence for a different framework. There are plenty of unexplained data already out there that point toward internet-like engineering.

The ignored or leftover data in computer metaphor–guided studies are often called *noise*. One class of noise is physiological noise. Individual neurons, it turns out, are not very self-consistent in their firing. Though some have claimed that noise of this kind is inevitable because the components of neurons are very small and therefore subject to random fluctuation, the real reason is that neurons are highly interconnected. It is the complex arrays of communication links—synapses—that make neurons difficult to predict.[8] We know this is true because if we grow a single neuron in a petri dish, we can accurately predict its firing in response to small zaps of current from an electrode. Firing means producing spikes, which are the stereotyped electrical signal produced in many neurons. In a petri dish, both the number of spikes a cell produces and the timing of those spikes can be predicted well using Hodgkin-Huxley–like models. It is only in the context of living brains that single neurons become unpredictable.

Say we show some glowing rectangles to an animal and record a visually responsive neuron in the cortex. Then we do the same thing again a few seconds later. We would be lucky if the spiking pattern the first time matched even half the pattern of spikes the second time. Such fundamental indeterminacy when passing ostensibly the same "message" suggests that this variability may not truly be noise after all.

The way noise is treated in neuroscience today is a direct result of the unspoken assumptions—the metaphor—we have adopted. Noise and randomness cannot be tolerated in computation and need to be explained away or ignored. A computer is an entirely deterministic machine. Given a set of starting conditions—say, multiplying two numbers together—any computer worthy of the name will always give the same answer. In fact, randomness is so alien to computers that it takes the most powerful quantum computers to make something approaching true randomness. And even then, more powerful computers can be built to de-randomize, again by rigidly implementing the laws of mathematics, as nonrandomly as possible.

Because randomness is unacceptable in a computer, it is treated the same way in many approaches in neuroscience. This treatment usually stems from the application of Claude Shannon's information theory in the brain. The idea of Shannon's theory—and its most central theoretical achievement—is to determine how noise can be minimized when electromagnetic signals are sent from one place to another. Information theory has been wildly successful in this role, and our modern technological society is utterly dependent on this insight.

If noise is bad for computers and the transmission of electro-magnetic signals, the computer metaphor implies that it is deleterious in brains as well. It would be illogical for the system to inject noise—unless it had goals other than optimal computing.

This brings us to the beginning of an answer to our first question.

• • •

Question 1: How does the brain deal with collisions between messages?

As we saw, noise has demonstrable benefits on the internet, in particular in back-off algorithms. These strategies provide insights into our noisy brains. Recall that, following a packet collision, simple back-off algorithms like those of ALOHANET force the two colliding messages to wait a random amount of time. This has the effect of injecting timing noise into the system. Senders are assigned a fixed probability—say, one out of three—and at each tick of the clock, they are able to resend their packets with exactly this likelihood. Each sender essentially picks a number from 1 to 3 out of a hat, and if the number was a 1, it is allowed to send its packet. This strategy results in very few collisions even though senders do the random number picking on their own. But it is a fairly conservative strategy. Lots of time is wasted by senders randomly choosing a 2 or a 3.

Today, the internet uses a more efficient approach based on what is called exponential back-off. Say two packets collide on their way to a receiver, causing both to be lost. The receiver then flips a coin to determine which sender will be allowed to try again first. The winner is notified and resends the lost packet.

A few microseconds later, the loser is invited to do the same. But if packets from the two senders keep colliding—for example, because the loser has sent a different packet at the same time the winner is resending the first—there is further pain. The loser of each subsequent coin flip needs to wait an increasing amount of time before being allowed to resend. The time delay is designed to increase exponentially. If two senders can't play nice, they are both punished at a sharply increasing rate. The upshot is that the chances are very good that a packet from either sender to the receiver will arrive with very little delay after a collision. Neither sender will have to wait very long before sending a packet again. The chances of having to wait a long time fall very quickly for longer and longer intervals.

Have brain networks found a way to exploit noise in similar ways as the internet? Networks of neurons certainly could inject noise in order to avoid collisions. Interestingly, the particular kind of randomness that is most useful for back-off algorithms on the internet today is very similar to the empirical pattern of variability in neural firing.

Classical neuroscience guided by the computer metaphor—which assumes that the spike rate is the only thing that matters—has attempted to explain away the noise in spike timing. The standard model for spike timing noise is called the *Poisson spiking model*.[9] After a spike occurs, a neuron waits a certain amount of time before spiking again, and this timing noise can be characterized in terms of a model based on the mathematics of what are called Poisson processes. The Poisson model predicts that the delay between successive spikes will follow an exponential distribution. In other words, the Poisson model says that the chances are very good that there will be a short delay between spikes, while the chances of a long delay decline rapidly for longer and longer delays.[10] This should sound familiar: an exponential distribution

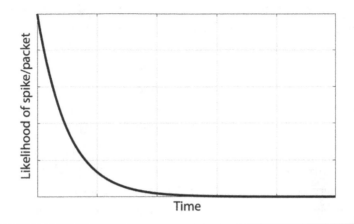

FIGURE 6.1 A falling exponential function describes the likelihood of a neuronal spike at a range of times after an initial spike. That is, the curve describes the distribution of interspike intervals. The same function also describes the likelihood of a message packet on the internet being resent after an initial collision of packets. In the case of neurons, the exponential dropoff is predicted by what is called a Poisson process model of neuronal biophysics, while on the internet the exponential dropoff is imposed directly by the network's back-off algorithm. In both cases, this behavior efficiently spreads network activity over time and can lead to fewer collisions. Image by Daniel Graham.

of delays is precisely the behavior engendered by the internet's back-off algorithms in order to avoid collisions (see figure 6.1).

The Poisson model is a good first approximation for characterizing spiking in lots of different types of neurons in the brain.[11] Much of the time, delays between spikes do indeed occur randomly in a way that follows an exponential distribution. But because the Poisson model is usually deployed in the context of the computer metaphor, the goal has been to explain how certain computations like multiplication can give rise to Poisson-like processes, and how timing noise could be corrected by subsequent computations.

In contrast, the internet metaphor sees the delays between spikes as a way of organizing message passing in the brain so that collisions are kept to a minimum. The brain could be utilizing the randomness of an exponential back-off algorithm to prevent collisions of messages, rather than finding ways to excise noise from neural computations. Avoiding collisions would be especially important on the small world–like network of the brain, where any neuron is very close on the network to practically any other neuron. Neurons could inject random temporal variation on purpose to offset signals. If the system's protocol were appropriately designed, adding timing noise would not necessarily lead to message corruption, which could in any case be corrected with other strategies of communication engineering such as the ack (discussed in the next section).

A collision avoidance strategy like this may help explain what we call noise in messages of many spikes. Under most applications of the computer metaphor, if the goal is to explain how many spikes we are likely to measure, then the error in our prediction doesn't need interpretation: it is just random, meaningless fluctuation, which can be corrected by other stages of computation. It should be noted that some schools of thought based on the computer metaphor do ascribe meaning to the variability in neural firing. There are numerous models of what are called spike timing codes.[12] In this view, the timing of spikes carries information over and above the rate of spiking. Spike timing may well be a component of a neural message: in a burst of spikes, perhaps the message's destination is encoded in spike timing. However, what I am proposing in relation to exponentially distributed spike timing is that the noise is indeed random, that it is there on purpose, and that it serves *communication* goals. The purpose is to manage the inevitable collisions of messages.

• • •

Question 2: How is reliable delivery of messages achieved?

The brain's strategies for achieving reliability might again look like noise to a physiologist working under the assumptions of the computer metaphor. But this apparent noise could serve functional purposes. Useful signals may appear to be noise because we assume they are related to computational goals rather than communication goals.

If we are recording from a neuron in the visual brain, we would normally assume that it is only "speaking" in response to a stimulus and is computing something about that stimulus in what it says. In other words, we would normally be looking for a relationship between some aspect of the stimulus (like the size or shape of a pattern of light in the world) and some number of spikes. How could we tell if the neuron were, for example, saying something about successful transmission instead?

In addition to timing noise in classical neurons, there is also what we might call anatomical noise. This kind of noise comes from neurons whose responses we simply ignore. It includes neurons in a given functional area like the visual system that don't reliably respond to changes in light on the retina but still fire on occasion. Based on by far the most comprehensive data to date—a massive study from the Allen Institute recording from 60,000 neurons in 243 mice—a class comprising a plurality of neurons in visual cortex were found to be "not reliably responsive to any of the stimuli [tested]."[13] In the cortex's auditory system, only around 10 percent of neurons respond to sound stimuli.[14] There are also many neurons in the brain that don't respond to any stimuli, be they visual stimuli or loud noises, a puff of air on the face, the smell of food, and so on. These neurons too may spike from time to time.[15]

Cells of these kinds are inconvenient for models of brain systems that assume optimality because they don't produce enough data to be modeled, or don't fit an existing model. They are typically ignored.[16] Until the early 2000s, the exclusion of these kinds of data was not even mentioned in most studies of single-neuron electrophysiology (and they are still sometimes omitted today). Yet these cells may hold a wealth of information about communication-related goals in the brain. They are what we referred to in chapter 3 as neural dark matter.[17]

Spikes from neural dark matter may serve as acks, letting senders know that a particular tranche of messages was received somewhere else on the network. These spikes would be very sparse. First, they would come in response to the firing of ordinary neurons, which are presumably the ones trying to get messages across. Since we know that most neurons, including putative senders, fire sparsely, the acks returned to them would also be sparse. Second, ack messages—which only need to say "got it"—can be very small, perhaps even a single spike, making them even sparser.

The extreme sparseness of many neurons in the brain may be why we have ignored them.[18] At least some of these neurons may be involved in acknowledging message receipt. After all, large-scale communication networks almost always have some strategy for verifying message delivery, even if it is a low-tech solution like saying "hello?" when we answer the phone.

The structure of the connectome may also provide a substrate for an ack-like system. The small world–like network topology of the brain means that the task of passing acks can be distributed to many cells, since there are usually multiple short paths between sender and receiver. Brain areas like the thalamus may also be specialized for acks. As we saw in chapter 4, the thalamus is part of the network backbone of the brain. Messages from the eyes and ears and other sensory inputs pass through the thalamus on the way to their respective areas of the cortex, and these cortical areas send messages back to the thalamus, which sends them on again to the cortex. Each of these systems has thousands or millions of connections to and from the thalamus, forming great loops in the network.

Perhaps these loops are in part designed to allow ack messages to pass from a receiver back to a sender. Finding evidence of acks in the brain requires being able to trace signals traveling across more than one synapse in the brain, which is rarely possible. But innovative research methods have shown that the thalamus is fully capable of sending messages in a loop—and fast.

American neurobiologists Farran Briggs and Martin Usrey showed in 2009 that signals can make a round trip from the thalamus to the cortex and back to the thalamus in as little as 9 milliseconds. In monkeys, Briggs and Usrey showed this by injecting a large, distinctive spike of current into the thalamus, then measuring when this spike arrived in the cortex and when it returned to the thalamus.[19] Since visual perception takes at a minimum 150 milliseconds, waiting 9 extra milliseconds for a return receipt wouldn't necessarily slow the whole process down very much. And the delay would be worth it if it promoted reliability. When a sending neuron doesn't receive a brief "I got it" from a downstream neuron in timely fashion, the sender may be triggered to resend the missing message. This could help ensure accuracy in our perceptions (and in motor plans, which also pass through the thalamus).

Thus, in the thalamus, we have a part of the connectome that takes in sensory information via connections from the sense organs, links on to the cortex, and then loops back to the thalamus. It also helps control motor plans. Both of these functions need to be reliable, with errors in message passing quickly corrected.

The network architecture of the thalamus is a central and defining feature of the brain for all mammals. Under the computer metaphor, the thalamus has been seen as serving to adjust the strength of signals on their way elsewhere. But it is not clear why this function cannot be performed locally in the thalamus, and without loops to and from the cortex, as noted in chapter 1. Under the internet metaphor, the looping architecture of the thalamus could have evolved in part as a strategy to support reliable delivery of messages across most parts of the cortex, using something like acks.

One intriguing possibility is that the reticular nucleus of the thalamus manages ack-like functions. The reticular nucleus is the hot-dog bun we saw in chapter 4 cradling the double hot dogs of the left and right thalamus. Neurons in this area exert inhibitory influence on messages passing from the thalamus to the cortex. By default, parts of the thalamus involved in vision, hearing, touch, and so on could assume that they should send their messages to their respective cortical targets repeatedly until told to stop. If a message arrives successfully in the right part of the cortex, an ack could be sent back to the thalamus by way of the reticular nucleus. This inhibitory signal would in effect tell the sender neurons in the thalamus that they need not resend their message. Thus, the return-receipt system wouldn't work in exactly the same way as acks or nacks on the internet, but the goal would be the same and the mechanism would be similar.

• • •

Question 3: How is flexible routing accomplished?

This is a large and critical question. The strategies employed by communication systems to achieve flexible delivery are at the core of their routing protocol; communication systems wouldn't be very useful if users had no way to choose whom they communicate with.

The internet metaphor provides us with both general strategies and specific solutions for how the brain can direct signals in a flexible way. The problem is to send messages to one place sometimes and to other places at other times. We will start by considering general strategies for flexibility and their relationship to human cognitive function. As we will see, highly flexible cognitive functions have long been recognized, and they are given a variety of descriptive terms. But so far human cognitive flexibility hasn't been strongly linked to flexible neural mechanisms for routing messages.

To remedy this situation, I will outline more than a dozen neural mechanisms that could help accomplish flexible routing in the mammalian brain. Most of these mechanisms have already been shown to be fully capable of flexibly directing signals in living brains, but few have been identified as routing mechanisms per se. The internet metaphor helps us see that these mechanisms could be part of a concerted, brain-wide effort to route messages flexibly. Some of these neural phenomena could play a supporting role in flexible routing by doing things like monitoring network status and finding efficient paths, again in ways that mimic the internet.

The main reason we need flexible routing is that flexibility is useful for lots of things in lots of species. Flexibility is especially important for humans. Flexible and willful cognition is the hallmark of our species. But the computer metaphor and the dominant theories in neuroscience don't allow for much flexibility. As we have seen, computers and their neural analogs are only useful if they do the same thing every time for a given input.

There are many names for the basic element missing from this picture: one could call it unpredictability. Cognitive scientist Philip Lieberman writes: "Unlike ants, frogs, sheep, dogs, monkeys or apes—pick any species other than *Homo sapiens*—our actions and thoughts are unpredictable."[20] He argues that our species' unpredictable behavior is a reflection of our powerfully flexible cognition.

Neuroscientists have invoked other terms to describe the power of our flexible brains. F. Gregory Ashby, in his classic 1952 book, *Design for a Brain*, identifies *adaptability* as the key design feature of brains.[21] Still others use have used the terms *intelligence* or *purpose* to describe the same quality. I prefer *flexibility* because it doesn't imply the disorder of unpredictability, the optimization of adaptability, or the value judgment of intelligence and purpose. Flexibility implies "good enough," and it is a general goal that can aid in many types of tasks, including tasks we don't typically regard as cognitive.

It is hard to overstate the importance of flexibility in human cognition. Because of its manifest importance, researchers have come up with numerous terms to describe it that often have overlapping meanings. One term is *cognitive control*, which computational neuroscientist Jeff Beck and colleagues define this way:

> Cognitive control describes the strategic guidance of behaviour in accordance with internal goals. A key feature of cognitive control is flexibility, that is, the continual adjustment of processing strategies in response to varying demands. For instance, we are able to mobilize and adaptively shift attention towards a particular task when we encounter difficulties, such as worsening weather conditions during a road trip.[22]

Cognitive control is considered one of two main features of human intelligence, along with semantic knowledge.[23] Scientists test cognitive control in the laboratory in various ways, such as by measuring how well people can switch between tasks. The tasks could be finding a specific color in a set of flashed symbols and then switching to finding a particular shape. The archetypal cognitive control task outside the laboratory is chess. Each turn involves mentally switching among our pieces to imagine possible moves and those of our opponent. It seems no coincidence that performance on lab tests of cognitive control and achievement in chess competition peak in the same age group, in the late teens and early twenties. Cognitive control is all about flexibility: changing our mental state in strategic ways to think about things in different ways at different times.[24]

Researchers such as neuropsychologist Peter Tse of Dartmouth College use the term *mental workspace* to describe a similar ability.[25] Tse emphasizes the volitional nature of cognitive flexibility. He quotes Albert Einstein, writing about his scientific process as involving "certain signs and more or less clear images which can be 'voluntarily' reproduced and combined."[26] Not only do we have lots of options for combining and remixing our stored knowledge, Tse argues, but we can also actively choose how and when to perform these mental operations. Our free will is strong and can be invoked seemingly at any time. Harking back to Descartes, this notion may seem a bit dualistic. The term *cognitive control* and the volitional nature of the mental workspace evoke a diligent worker at a desk inside our heads, manipulating symbols or chess pieces. We know there is no little person inside us performing these operations. But the mental workspace concept does point in the direction of a flexible communication system underlying our flexible cognitive functions.

Psychologist Fiery Cushman of Harvard University uses the term *representational exchange* for much the same idea. It has the advantage of lacking dualism. Cushman defines representational exchange as the "transfer [of] information between many different kinds of psychological representations that guide our behavior."[27] This concept helps highlight the intermodal or interoperable nature of the brain. Representational exchange gets us still closer to thinking in terms of communication protocol. Yet this conception is still centered on the "content" of cognition, and not on how flexible exchange could be implemented on a large communication network.

Suffice it to say, cognitive scientists are forthright in highlighting the importance of flexible interchange of information in the human brain. But I find it curious that modern neuroscience has so far not seen the relevance of an internet-like routing protocol in brain networks to support flexible cognition. Indeed, cognition has even been compared to the PageRank algorithm of the Google search engine, but without positing the communication infrastructure that allows internet search.[28] Some have compared cognition to "cloud computing," but again without reference to the network protocol that makes this possible.[29]

The internet metaphor doesn't fully explain flexible cognition, but it gives us a useful reference point, and one that is not itself dualistic. The internet achieves control not through global overlordship, but through distributed rules. It is built on sharing. Each neighborhood exchanges information of several kinds. It's not just "content," either: there are also acks, as well as network status signals, and lists of efficient routes to more distant parts of the network. The system as a whole wouldn't work without these local, distributed exchanges. Above all, message passing on the internet is designed around flexibility.

As we saw in chapter 1, decision-making is a cognitive function that requires flexible message passing. A further characteristic of cognition is the ability to learn and remember. We can put these aspects of cognition into the new context of the internet metaphor as well. When we learn, we are in some ways removing flexibility. We are limiting the number of possible states to only those that relate to a particular action or bit of knowledge. This requires memory in some form. But our learned thoughts and actions can still be fine-tuned, and just as importantly, they can be altered as new needs arise.

Drawing on the internet metaphor, we can view learning as a kind of best route for messages to travel *for the moment*. Given current needs and network conditions, the best way to carry out a task is the one we have learned. On the internet, though each packet composing a message can in theory take a different route to the destination, most of the time the packets for a given message all take the same route. The internet sticks with what works for the moment—thereby diminishing flexibility—but it can readily learn a different route if the need arises. The brain may behave in the same way.

Moreover, change is possible. For example, if a message on the internet experiences many collisions along a particular route—which result in sharply escalating wait times imposed by back-off algorithms—it may be better for the sender to find a new route. In this way, the network learns something new, all the while retaining flexibility.

The internet metaphor gives a very different conception of learning and memory compared to the computer metaphor, where learning is equated with hard drive–like memory. More specifically, the "memory trace" in this context is instantiated in the fixed synaptic strength of a population of neurons. This is what is called Hebbian learning. In contrast, the internet metaphor sees the route connecting assemblies of neurons as constituting the learned knowledge itself. In short, an internet-like brain learns when it finds a good route.

In the brain, we can also see the flexibility of learned information in how goals are executed. Neuroscientists who study motor systems use the term *multirealizability* for the fact that there are usually many ways to achieve the same outcome of action. For example, we can write our name with a pen in our writing hand, or with our opposite hand, or even the mouth and feet. Our name may not look exactly the same each time, but the same representational knowledge—the letters of our name—will be generated. Crucially, this doesn't require new learning—we can accomplish the basic task without any practice. In other words, when we learn how to write our name with our writing hand, we also gain the ability to reach this goal through other means. This might come in handy if we injure our writing hand. Thus, multirealizability illustrates the idea that the path of the messages is what constitutes learning, rather than a specific set of computations or stored data. The task is to get the message to the right destination at the right time, rather than to compute a particular set of muscle trajectories. This is what routing is all about.[30]

Viewed through the lens of the internet metaphor, cognition and behavior are highly flexible, and even when we reduce flexibility through learning, we still have alternate paths available to execute learned behavior, which can be called on if needed. But how flexible are real neuronal networks?

At a neural level, we know that the brain has a remarkable capacity for performing a function with a different set of neurons when part of the network is damaged. If there is damage to a part of the brain that performs a particular function, the network can redirect the flow of information and achieve much the same outcome as before. We know this is not because the network builds new connections around the damaged parts, but rather because it uses connections that were there all along.

This form of flexibility is seen most clearly in sensory systems. Consider again an example from touch perception we introduced in chapter 1. In parts of the brain that monitor our touch sensations, each little chunk of brain is responsible for analyzing signals from a particular patch of skin. When signals from a patch of skin are blocked or cut off, there is a piece of brain that is no longer being very useful, since it has no "content" to work with from the senses. But the idle piece of brain quickly takes on other jobs. This is called functional reallocation.

Distinct patches of neurons in the cortex track changes in pressure on each finger. We can block electrical signals from a finger at the elbow by injecting an anesthetic to a specific bundle of axons on their way to the cortex. When neuroscientist Edward Taub and colleagues did this to nerves from a monkey's thumb and first two fingers, they found that cortical neurons that had previously responded to these fingers now responded to the ring finger and pinky (and also to touch on the lip, whose cortical territory borders that of the thumb). Rerouting happened very quickly, within an hour of injecting the anesthetic.[31] This is far too little time for new axons to grow from the pinky pathways to the thumb-and-finger pathways.

In the visual system, rerouting can happen even faster. As we saw in chapter 3, each chunk of the retina is mapped to a corresponding chunk of the visual cortex. It turns out that when a part of the retina is damaged, its corresponding cortical territory starts responding to neighboring parts of the retina within minutes.[32]

Functional reallocation is one phenomenon where routing strategies used in engineered networks have already been proposed as useful comparisons for brains, thanks to Australian communication-engineer-turned-neuroscientist Andrew Zalesky. The idea is that both the brain and data networks could use something called deflection routing to redirect signals after blockage.[33] Deflection routing—a variant of the hot potato—is a mostly theoretical engineering solution today, but engineers are interested in it because it could help underpin optical communication networks. Essentially, if a message encounters a blockage on its route, it is deflected in a different direction and tries to make its way on to its destination over a less-than-optimal route. The strategy seems most useful for messages that travel very fast, like light pulses, but also for systems that can't store messages en route. In communication networks, deflection routing requires that routers know in advance about a variety of good routes. If functional reallocation in the brain resembles deflection routing, then it shows more than just a simple avoidance of the problem area or damage. It shows us that robustness and flexibility are fundamental to the system.

We can build on Zalesky's basic insight of connecting communication protocol with brain function if we think about the requirements and goals of functional reallocation. It turns out that the brain's solutions correspond nicely with those of the internet. They have to do with the rerouting of messages.

First, the speed with which rerouting occurs implies that the brain already knows about alternate routes. The network may have used these alternate routes from time to time before the damage occurred. Likewise, routers on the internet always have an up-to-date map of their network neighborhood, thanks to keep-alive messages and the sharing of routing tables. Even if alternate routes are not used very often before an interruption, each router always has fallback options available any time the need arises.

In neurons, spontaneous activity may serve to inform neighbors that a given connection or route is "alive." All neurons that produce spikes in response to inputs also produce spikes spontaneously, seemingly at random times. Spontaneous activity is yet another source of noise in most measurements of the brain. But perhaps spontaneous firing is not useless noise. Spontaneous spikes could be a kind of keep-alive message informing other neurons in the neighborhood that a particular link is still viable.

The second correspondence between the brain and internet highlighted by functional reallocation is that the networks are fully interoperable. The brain doesn't really care what the signals represent. Information about what our thumb feels and what our pinky feels is interoperable.

Interoperability has been illustrated in a beautiful series of experiments by neuroscientist Mriganka Sur of MIT. Working with newborn ferrets, axons from the eyes are detached from their usual destination in the visual thalamus and "sewn" onto the part of the thalamus that normally gets inputs originating from the ears. The axons reestablish connections with auditory regions within a few months, at which point the animals' *auditory* brain responds to light in much the same way as it does in the *visual* brain of normal ferrets. The rewired animals can even see well enough with their auditory brains that they can make basic visual distinctions, such as choosing a particular visual pattern associated with reward.[34] So, just as on the internet, the network seems to package different kinds of information in interchangeable envelopes—perhaps we could call them packets. Such information in the brain could be flexibly utilized by different "applications," depending on the context of the messages' arrival.

Functional reallocation also points to a more general phenomenon that highlights a correspondence between rerouting in the brain and in the internet: in both cases it is wise to spread message across many routes over time. While many neurons in the brain are rarely if ever active, the brain nonetheless abhors a vacuum. When big chunks of healthy neurons stop receiving the inputs they're used to, the brain works hard to give those neurons something else to do.

Moreover, neurons that we use rarely now might come in handy if we have an injury and our brain needs to find a new route for some messages. Sparsely firing neurons could provide surge capacity, just as sparsely active routers do on the internet. Unlike many other networks such as supply chains, the internet is designed to handle huge surges in activity. Most of the time, it operates far below capacity. This is only possible because of the flexible, distributed nature of the network and the fundamental goal of robustness. Perhaps the brain's normally sparse operation is also a reflection of surge capacity. If we have lots of neuronal routers, and each one has access to many routes across the whole network, we can redistribute signals in such a way that most neurons are used only in uncommon bursts.

These qualities of brain networks are not suggested by the computer metaphor but are central to the internet metaphor. Indeed, the brain-as-optimal-computer framework holds that the brain seeks to operate at close to theoretical limits, like a just-in-time supply chain. The success of the internet—and the breakdown of many supply chains—during the coronavirus pandemic illustrates why brains may require robustness too.

•   •   •

So the routing of information in the brain has correspondences with routing on the internet, and these ideas help us reconceptualize many aspects of cognition. We can now start to make nuts-and-bolts proposals about how the brain, as part of its normal operation, gets information to one place sometimes and to another place at other times using neurons. Classically, this capability is outside the realm of possibility. Though there are exceptions,[35] most neuroscientists believe that a neuron is bound to add up signals at its inputs and pass along the results of this computation to all other neurons joined via synapses to its axon. It can't choose to direct messages to different places at different times. But today there is a wealth of evidence that neurons can do just this.

Of course, we still have to make this new picture accord with what we know about brain computation. The goal is not to replace existing models, which are well supported in many cases, but rather to examine a different aspect of the system. A similar situation exists in modern physics: the quantum picture of fuzzy elementary particles must reduce to the familiar classical picture at macroscopic scales and velocities. Likewise, the internet metaphor shouldn't contradict neurobiology, and indeed none of the mechanisms I will propose are ruled out by existing neurobiological understanding.

It is certainly true that single neurons can often be activated by any of several inputs and that they are generally more likely to fire when they get more excitatory inputs (at least when they are alone in a petri dish). This is the integrate-and-fire model of neurons, which suggests that neurons usually perform a summation of many signals before firing. It's the archetypal computation in the brain, which we have seen in the models of McCulloch and Pitts and Hodgkin and Huxley. But if the brain mostly fires sparsely, the integrate-and-fire model may not be the only or most important scheme for information processing. The issue is that the chances of coincident excitation may not be very high in most parts of the network.[36] The living brain could instead generally aim to have just a few neurons sending signals to a given neuron. In this view, the branching input and output networks of single neurons serve mainly to allow many possible inputs and outputs, rather than to concentrate and sum together a lot of activity at each neuron.

Candidate strategies for flexible neural routing exist at three general scales: the single neuron, the cell assembly (a handful to thousands of neurons), and the brain region (thousands to millions of neurons).

At the smallest scale, we have the process of routing in single neurons. As it happens, there has long been a strand of theoretical neuroscience focused on the selective distribution of signals by single neurons. One leading proponent of this school of thought was Alwyn Scott, an unorthodox American mathematician whose well-supported models of neurons do not fit with the dominant paradigm. Defying the standard model, he was especially forthright in his insistence that neurons could distribute signals to one output sometimes and to another output at other times.

One reason neuroscientists have historically doubted the possibility of flexible routing in single neurons is that in most neurons, the part of the cell where the axon emerges from the cell body represents a bottleneck with only one path. But Scott had a solution. In 1977, he constructed detailed models in which, as he put it, "a real axon has the ability to translate a time code on the trunk of the axonal tree (embodied in the temporal intervals between impulses) into a space code at the twigs of the tree (embodied in the particular twigs that are excited)."[37]

Scott built on the earlier work of many neurophysiologists, including David Van Essen, and especially Stephen Waxman. Waxman coined the notion of the *multiplex neuron*, by which he meant that neurons could send signals from some parts of the axonal tree sometimes and from other parts of the tree at other times. Waxman identified four zones in which a neuron could modify signals in order to pass them to different outputs (figure 6.2).

The first two zones, at the dendrites and cell body of a neuron, can be modified by many external inputs. As Scott later argued, this has the effect of transforming signals over time, which, combined with the asymmetric dynamics of the neuron, can allow the whole cell to send messages to different outputs at different times despite the bottleneck of the axonal base.

The axon itself is the third zone. It often has several branches and offshoots that go in many different directions. Waxman proposed that the branching patterns could perform the "steering" of information, an idea taken further by Scott.[38]

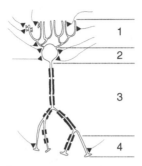

FIGURE 6.2 The multiplex neuron. Going against the orthodox notions of neuronal function of his time, Stephen Waxman in 1972 identified four zones in a neuron that could be capable of directing signals to different destinations at different times. Zone I is the dendrites, which, by virtue of their branching structure, could select between different signals incoming from the synapse. The cell body (soma) also receives signals via synapses from other cells. It constitutes Zone II, which also influences the dynamics of spike generation. Zone III is the axon, which, like Zone I, may have several branches capable of directing signals, but in this case could select outputs dynamically. The axon terminals (Zone IV) can also control signal flow by selectively gating outputs to the synapse. Since Waxman's time, these kinds of functions have largely been confirmed and in some cases extended. Image by Daniel Graham. Adapted from: Stephen G. Waxman, "Regional Differentiation of the Axon: A Review with Special Reference to the Concept of the Multiplex Neuron," *Brain Research* 47, no. 2 (1972): 284.

Axons are also subject to interactions imposed by other neurons. So-called axoaxonic synapses pass neurotransmitters directly to the axon from other neurons, which can enhance or block transmission along particular branches. It is even possible for axons to influence each other without exchanging neurotransmitters, through electrodynamic interactions, again potentially having the effect of steering output signals to different receivers at different times.[39]

The fourth zone where routing can occur is at the axon terminals. Here, the complex dynamics of neurotransmitter release, as well as the opportunity for influence from the axons of other neurons, provide many possible mechanisms for flexible routing. A given terminal could make message passing across its synapse more or less likely in a dynamic, selective way.[40]

Though they are hardly mainstream, the theoretical programs of Scott and Waxman are now being rediscovered by modern connectomics researchers.[41] Their ideas fit well with the internet metaphor. It begs the question of whether these ideas would have been more accepted in their time if they had had the advantage of a more detailed technological metaphor. The term *multiplex neuron* does not fully capture the flexible behavior of neurons that Waxman was trying to describe. Scott made the analogy of neurons being like "chips" in a computer that perform more complex functions than a single gate and that are capable of directing signals to more than one destination.[42] But routers do much more than that: they also correct errors and monitor network traffic and status. Devices that perform flexible routing were little known outside computer engineering in the 1970s, so the metaphors Scott and Waxman used were less precise than they can be today in relation to the modern internet.

With the benefit of several decades of further neurobiological discovery, we can look in more detail at some of the ways that single neurons could direct the flow of messages in the brain. Consider first the synapse. There are numerous ways the synapse could serve as a router of information. First, the biochemistry of the synapse is extremely complicated. The standard picture is that a spike in voltage arrives at the end of the axon, which triggers a series of reactions in lockstep, leading to the release of neurotransmitters and ultimately to the reception of those neurotransmitters in the dendrites of the receiver neuron.

In reality, there are a dizzying array of internal chemical interactions at every stage of this process that can potentially change the outcome. Gene transcription adds an additional layer of complexity. Without wading too far into these waters, I mainly want to point out that many possible mechanisms exist that could allow flexible routing within a single neuron. For example, the flow of calcium ions into the neuron at the axon terminal, which can trigger neurotransmitter release, is a rich source of dynamics. Calcium flows can alter and be altered by many other chemicals, including proteins, phosphates, and enzymes. Some have even suggested that quantum interactions could affect synaptic dynamics in meaningful ways.[43] In any case, no current evidence rules out the possibility that axon terminals could support a system capable of passing messages to different receivers at different times.

However, neurons don't pass messages all by themselves. Each neuron is surrounded by accessory cells. These are called glia, or helper cells that support neurons by providing nutrients, insulating axons, and performing other vital functions. The synapse is increasingly seen as being supported by glia as well. Some have called this notion the *tripartite synapse*: rather than being composed of two elements—the axon terminal on one side that releases neurotransmitters, and the dendrite on the other side that receives neurotransmitters—the synapse now has a third component consisting of glial appendages.

Certain features of glial interactions at the synapse are particularly well suited to the demands of flexible routing. In 2007, German neuroscientist Christoph von der Malsburg and colleagues made a detailed argument along these lines.[44] As they point out, the glia of the tripartite synapse—cells called astrocytes—are particularly good candidates for performing routing.[45]

First, astrocytes have receptors for some neurotransmitters, meaning they can sense information transfer between neurons. Second, they can modulate the passage of neurotransmitters by releasing "gliotransmitters," chemicals that serve similar functions as neurotransmitters but are released by glia. Some gliotrans-mitters are chemically identical to neurotransmitters such as glutamate, while others are amino acids, lipids, and phosphate complexes, which can affect neurotransmitters and neurons in a variety of ways. At the synapse, gliotransmitters can modulate both the sender and the receiver of neurotransmitter messages (i.e., both the pre- and post-synaptic neurons). Third, a single astrocyte can participate at tens of thousands of synapses. This means that astrocytes can potentially help direct many messages simultaneously. Finally, there is now evidence that glia have access to sensory signals and that they can use these signals to control messages sent by sensory neurons. Researchers have recently shown that astrocytes are capable of changing the activity patterns of neurons that detect air pressure changes in the mouth. In particular, glia cause the neurons they interact with to respond in different rhythmic ways depending on the context of airway stimulation (chewing versus breathing, for example).[46]

There are other potential mechanisms for flexible routing at the single neuron level. One candidate brings us back to Ramón y Cajal, who was among the first to study a group of neurons called *pericellular nests*. Ramón y Cajal's experiments led to some of his most curious and beautiful anatomical drawings (see figure 6.3). Pericellular nest cells have been compared to balls of yarn surrounding other neurons.[47] They seem capable of controlling the outputs of the enclosed neuron. Pericellular nest neurons occur in many parts of the brain in vertebrates. Related neurons called basket cells have a similar shape and are more numerous. In mammals, these cells often enclose the workhorse pyramidal cells of the visual cortex. Physiologists have found evidence that these kinds of cells "establish multiple synaptic contacts with the neurons they seem to select as postsynaptic targets."[48] Nest and basket cells participate intimately in the ability

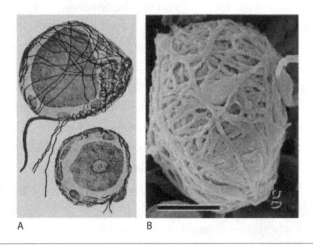

A   B

FIGURE 6.3 A: Detail of drawing of pericellular nest cells by Ramón y Cajal (1909). B: Electron micrograph of pericellular nest cell in the spinal cord of the rabbit. Image Sources: (A) Ramón y Cajal S. 1909. *Histologie du systeme nerveux de l'homme et des vertebras.* Tome I, II. Paris: Maloine. (B) Matsuda, Seiji, Naoto Kobayashi, Takehiro Terashita, Tetsuya Shimokawa, Kazuhiro Shigemoto, Katsumi Mominoki, Hiroyuki Wakisaka et al. "Phylogenetic Investigation of Dogiel's Pericellular Nests and Cajal's Initial Glomeruli in the Dorsal Root Ganglion," *Journal of Comparative Neurology* 491, no. 3 (2005): 240.

of a neuron to communicate, and they seem capable of doing so flexibly. Basket cells also enclose motor neurons and neurons in other parts of the nervous system. Moreover, these cells are more numerous and elaborated in larger animals, where flexible behavior tends to be more important. Thus, there appears to be more than one class of neurons whose job is not to respond to stimuli but rather to direct and control the flow of signals from the neurons enclosed within their ball of yarn.

• • •

We can widen our focus beyond the single neuron and its accessories to assemblies of neurons. This level of analysis builds on much existing work in neurobiology, but the relevant findings have only rarely been seen as indicative of routing.

Mainstream theories do permit *gating* in neural assemblies, which is where one set of neurons acts as a gate on another set of neurons in a circuit.[49] Gate neurons either allow or block signals from the senders, controlling whether senders are able to pass information elsewhere or not. In some circuits, the gate can be opened by the presence of activity in the gating neurons, and in other circuits the gate is opened by lack of activity. Systems in many parts of the nervous system are thought of in these terms, with the classic example being the gating of pain perception. When a pain signal is generated in our limb—say, we accidentally touched a hot frying pan—the signal has to travel up the spinal cord to our brain before we can consciously perceive the pain. To get to the brain, signals from pain receptors pass through a neural gate in the spinal cord and travel on to the cortex. If gating neurons are activated by something else, such as mild electrical stimulation in an area near the injury, the gate will be partially closed. Consequently, fewer messages of pain will be transmitted up the spinal cord to the brain, and the injury won't hurt as badly as it would if the gate were open.

Gating of this kind is undoubtedly useful to brains, and it probably forms a component of the global routing scheme used in the brain. But it can do more. Two researchers of the retina have shown how neurons can implement the selective routing of messages using gates. Tim Gollisch and Markus Meister have highlighted the fact that gates in the retina can choose to pass sensory messages to the brain from one group of sender cells or from a complementary group of sender cells (figure 6.4). The gate neurons' choice depends on aspects of the sensory environment separate from the ones being measured by the two groups of senders.

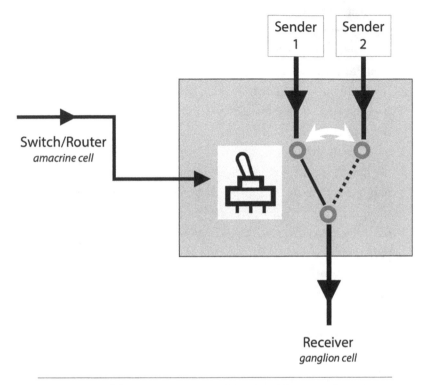

FIGURE 6.4 Selective gating in neural circuits of the retina. Amacrine cells are a class of neurons in the retina that can selectively route the flow of information. Tim Gollisch and Markus Meister have shown that certain patterns of motion in the visual image cause amacrine cells to switch between different senders of signals, with the selected output being passed to destinations in the brain. Image by Daniel Graham. Switch icon by Arthur Shlain (CC-BY), https://thenounproject.com/term /toggle-switch/731364/. Adapted from Tim Gollisch and Markus Meister, "Eye Smarter Than Scientists Believed: Neural Computations in Circuits of the Retina," *Neuron* 65, no. 2 (2010): 153.

Recall that neurons in the retina called ganglion cells respond to particular patterns of light such as what I termed donuts and donut holes. Gollisch and Meister have shown that rapid motion in peripheral vision can trigger a gate composed of neurons called amacrine cells, which have access to motion information. Changing the state of the gate changes the inputs to the ganglion cells between two populations of senders (bipolar cells, which connect to photoreceptors). This switch allows the ganglion cells to change from looking for donuts to looking for donut holes. Gollisch and Meister were the first to use the term *routing* to describe the process, but this terminology hasn't caught on yet.[50]

It is somewhat surprising that routing is found in the retina because the retina is mostly a one-way street. Signals travel from photoreceptors, where light information enters the retina, on to the optic nerve (composed of ganglion cell axons), where spikes leave the retina and go to the thalamus and then the cortex. The retina is a feed-forward part of the nervous system not normally associated with flexibility. If routing is useful and feasible in an area like the retina, it is also likely to be so in the far more flexible cortex. Computational neuroscientists Thomas Gisiger and Mounir Boukadoum have recently provided elaborate models of how gates can generate flexible routing in many parts of the cortex.[51] But this notion remains relatively little explored.

To the extent that we lack more evidence of routing at the level of cell assemblies in the cortex, it may be due to the fact that interpreting the "content" of neural signals in the cortex is much harder than interpreting the content of signals in areas like the retina that are only a synapse or two away from the physical environment. If we can't separate signals related to content from signals related to routing, it will be hard to find evidence of routing. This is a challenge for most classical approaches in neuroscience.[52] But if we assume that dynamic routing is taking place, we stand a better chance of identifying its implementation in neural assemblies. The internet metaphor provides a wealth of ready analogies for the brain's strategies.

From the level of assemblies of neurons, we can zoom out to larger brain regions and their intricate patterns of connection and activity as a mechanism for flexible routing. Certain brain regions seem particularly well suited to the task of directing the flow of messages due to their architecture. For some regions, we have suggestive evidence from living animals of a flexible routing function, but for most brain parts, this function is at present only implied by anatomy. The issue, once again, is that it is very difficult to record neural activity in more than a handful of neurons. The resolution needed to trace message flow in millions of neurons is not achievable with brain imaging. Thus, the schemes I will describe for flexible routing mechanisms in larger brain regions are not as specific as those for single neurons or local circuits. Nevertheless, they fit with the larger need to route signals dynamically across the whole brain, and they are not ruled out by neurophysiology.

Starting at the "bottom" of the brain, in the brainstem, brain architecture suggests the possibility of flexible routing. A collection of regions called the *reticular formation* is the most obvious candidate. The name *reticular* refers to the net-like appearance of neurons in this collection of brain areas. Several of the nuclei that constitute the reticular formation make connections with almost every other part of the brain, as shown in recent connectomics studies. Using tracer molecules, researchers have found that areas like the raphe nuclei of the reticular formation receive inputs arising from touch on the skin, pressure in the gut, taste, sound, reward systems, many parts of the thalamus, eye movements, motor planning, motor execution, emotion, and lots of other systems.[53] The outputs of the raphe nuclei are likewise very widespread, reaching the frontal lobe of the cortex, memory areas, and numerous areas involved in planning, emotion, and other tasks.[54] The raphe nuclei as well as other components of the reticular formation, including the locus coeruleus, are critical for awareness and arousal. These areas are described, following the computer metaphor, as a clock or "pacemaker" for much of the brain since they are involved in sleep-wake cycles.[55] In other words, parts of the reticular formation are equated with the clock of a computer that keeps components computing in lockstep.

But perhaps a better analogy for their function is as exchange points in the backbone of the internet, where many subnetworks share connections. Unlike the metaphor of a computer clock, the notion of an exchange point implies flexible interchange of many different types of messages on a somewhat equal footing.

While full maps of connectivity of the brainstem are just starting to be worked out, more is known about the architecture of another component of the brain's network backbone: our old friend the thalamus. Here we have more anatomical and physiological evidence of how traffic can be controlled at the neural level.

The thalamus sits above the brainstem, surrounded by the cortex. It receives inputs from the sensory surface (eyes, ears, tongue, skin) and also forms connective loops to and from most parts of the cortex, as we have seen. These loops can convey messages very quickly and to many parts of the brain. Each time a message enters or leaves the thalamus on a loop, it passes through the reticular nucleus (not to be confused with the reticular formation, though the etymology has the same implication of a net-like structure). As noted earlier in this chapter, the reticular nucleus may be involved in an acknowledgment system. But it may do more.

Because reticular nucleus neurons can modulate most inputs and outputs, they are well positioned to direct message traffic. The connecting axons of reticular nucleus neurons have branches that send signals throughout the dense network of the reticular nucleus, and each wire also receives signals through long dendrites from other reticular axons. The particular types of connections of reticular nucleus neurons are also worth noting. These neurons have high interconnectivity via dendrite-to-dendrite and axon-to-axon chemical signaling, along with classical axon-to-dendrite synapses. They are also connected via extensive gap junctions, which provide fast exchange of chemical information.

So we have the scaffolding of a densely connected network using many parallel communication signals that is pierced by wires to and from the cortex. The traffic over these wires is thereby influenced by the network interactions in the scaffolding. This system would seem suited to serving as a routing system for signals going to or from the thalamus. The reticular nucleus of the thalamus could also serve as a buffer for messages. Since congestion and collisions are a fact of life in communication systems, it may be necessary to store messages for a period of brief time until they can be transmitted again. Consistent with this idea is the fact that thalamic reticular neurons seem to possess their own special code of electrical signals. Along with regular spikes, they produce spikelets, a lower-amplitude, higher-frequency form of excitation of the cell.[56] Perhaps spikelets are involved in a form of, which is to say caching a small amount of information in memory.

In this view, once signals reach the cortex, they can move laterally within the cortex and make a return trip to a different part of the thalamus, again with the possibility of routing by reticular nucleus neurons. Messages could explore the entire cortical network in just a few hops. This hypothetical scheme would potentially be fast—and reliable. As described in chapter 4, the looped architecture of connections between the thalamus and the cortex means that ack-like messages can be returned on the same system.

Another brain area that is a potential exchange point is located near the thalamus, called the claustrum, which is a part of the cortex (figure 6.5). Neuroscientist Christof Koch and Francis Crick, the Nobel Prize–winning codiscoverer of the structure of DNA, have argued that this area could be a critical locus of consciousness because of its extraordinary interconnectivity.[57] The

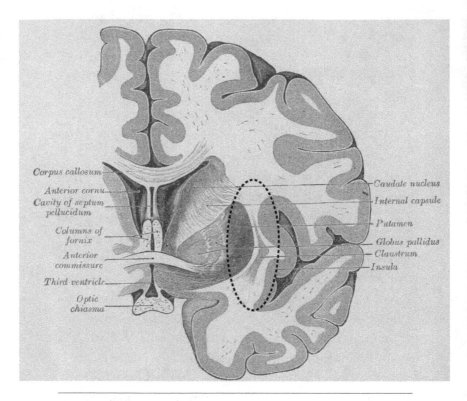

FIGURE 6.5 The brain's claustrum, viewed in coronal section (a slice from the top of the head roughly through both temples). The claustrum, a part of the cortex, provides high levels of interconnection among a host of cortical areas. Under the internet metaphor, this area could serve as a peering link that allows shortcuts between subnetworks of the brain. Adapted by Daniel Graham. Image source: https://archive.org/details /anatomyofhumanbo1918gray/page/836/mode/1up

claustrum sits amid millions of axons crisscrossing the cortex. As in the thalamus, neurons often form loops from the cortex to the claustrum and back to the same area of the cortex. The claustrum's job seems to be to share sensory information in real time among a host of areas. Koch and his team at the Allen Institute in Seattle have shown with tracer molecules that a neuron in the mouse claustrum has an axon that wraps around the entire brain, forming what Koch has called a "crown of thorns."[58] Thus, the claustrum seems capable of exchanging signals from myriad senders and receivers, and it also possesses long connective wires that span much of the brain.

Considering the thalamus and claustrum together, we have in the former an area that provides dense interchange of multimodal data. In the claustrum, we might have something closer to "peering" between subnetworks: it could provide shortcuts between cortical areas using long-distance links that bypass the thalamus. These two networks seem to provide efficient paths between widely dispersed populations of neurons, which could support flexible, reliable, and fast delivery of messages to all parts of the network.

There is tantalizing evidence from living human brains that large regions can selectively route messages to different destinations as well. Because this evidence comes from imaging, it is harder to point to specific mechanisms or structures underlying flexible routing, since imaging lacks the necessary resolution in time and space. But the evidence is nonetheless useful.

Michael Cole of Washington University in St. Louis and colleagues have shown how the frontal and parietal lobes of the human cortex can flexibly interchange information from different modalities, as well as attention, motor systems, and core functions.[59] Because frontal and parietal areas are associated with planning, volition, and higher cognitive functions, they could accomplish the flexible routing necessary for mental workspaces or representational exchange.

In research using a similar functional imaging approach, Columbia University neuroscientist Daphna Shohamy, working with Danielle Bassett of the University of Pennsylvania and colleagues, has found evidence for flexible routing during a learning task in an area of the brain called the striatum. The striatum, a group of regions that in part resembles a pair of tadpoles, each with its tail wrapped around itself, is thought of as performing functions such as reward, reinforcement, and motivation, along with initiating or inhibiting motor responses. But how it coordinates activity in the cortex is not understood.

Shohamy's team asked people to learn arbitrary associations between pictures of objects based on feedback (you chose correctly or you chose incorrectly). In the fMRI scanner, they showed that during this task three nuclei of the striatum can alter their outputs so that very different motifs of the cortex are engaged. The different motifs were associated with different aspects of learning, such as the rate of learning.[60] Based on its connectivity, the striatum is known to be a multimodal area that has been described as a network hub.[61] From the work of Shohamy and colleagues, we can see that the striatum seems capable of selectively directing traffic to widely dispersed parts of the cortex via its extensive network.

The internet metaphor's framework for understanding activity in large brain regions contrasts with the computer metaphor framework, which tends to see each brain region as performing a particular, fixed computation. This computer metaphor thinking is particularly common—and unexamined—in brain imaging research. Yet it has long been recognized that brain regions can play different roles at different times.[62]

Beyond regions like the frontal and parietal lobes of the cortex as well as the striatum, flexible routing may in fact be a fundamental design goal of the entire cortex. The cortex may possess some neurons whose job is to pass meaningful messages from one place to another, and other neurons—which fire more sparsely—whose job is to serve as auxiliaries to message passing, possibly doing things like routing and acknowledgment.

From detailed single-neuron physiology studies, there is increasing evidence that cortical cells generally perform either primary or auxiliary functions to control how the cortex communicates within itself and with the thalamus. Neurobiologist Murray Sherman of the University of Chicago, a pioneer in the study of the mammalian thalamus, has suggested that cortical neurons fall into two categories: drivers and modulators (figure 6.6).[63] The job of drivers is to pass primary messages from the cortex to the thalamus. The thalamus also receives primary messages from the eyes, ears, and other sense organs, but these are a surprisingly small proportion of inputs to the thalamus—only about 5 percent of the total, as we saw in chapter 4. The job of modulators—which are likely far more numerous than drivers—is to modify the messages from drivers. Typically, the driver-and-modulator picture is meant to subserve computation, in the sense of finding the right way to mathematically transform a given signal. Modulators could instead be seen as analogous to routers on the internet. Modulator neurons could deliver acks, as routers do, and they could also help send messages to different places at different times.

The brain's communication protocol at the largest scales may also perform network status functions that support flexible routing. This could include monitoring network link availability and traffic, as well as efficient pathfinding and addressing.

The *default mode network* comprises an array of areas in the human cortex, as well as in the hippocampus, an area that participates in memory and the understanding of 3D space. The most notable feature of the default mode network is that populations of neurons in this network are active when we are doing nothing in particular. This background activity spreads across many areas but doesn't increase much when we start doing something more taxing, like solving a math problem.

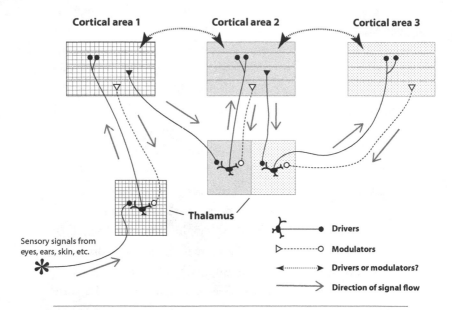

FIGURE 6.6 A schematic picture of information flow among regions of the cortex and between the cortex and the thalamus. S. Murray Sherman and colleagues have proposed that two general groups of neurons control the exchange of signals: drivers (solid lines) and modulators (dashed lines). Arrows indicate the direction of signal flow; some interconnections could be drivers or modulators (dotted lines). Signals arriving from sensory receptors (skin, eyes, ears, etc.) all pass through the thalamus, which is connected via drivers to respective primary sensory areas of the cortex. However, primary areas also send messages back to the thalamic area of origin of these signals by way of modulator neurons. Primary cortical areas also send messages to other areas of the thalamus and to other regions of the cortex. Each of these cortical areas likewise sends and receives driver signals from the thalamus, and also sends modulator signals to the thalamus.

Adapted from Iva Reichova and S. Murray Sherman. "Somatosensory Corticothalamic Projections: Distinguishing Drivers from Modulators," *Journal of Neurophysiology* 92, no. 4 (2004): 2195.

Though it is postulated as a substrate for thinking and rumination in humans, the default mode network could also be conceptualized as a system that monitors network status across the whole cortex or brain. Unfortunately, we don't know how sparse the firing of neurons in the default mode network might be, or much of anything about information coding at the individual neuron level as pertains to sharing information across this network. The problem is again that the network, which comprises numerous large regions, can only be studied at present using brain imaging, which lacks the necessary resolution.

But it seems clear that most messages passed across the default mode network are not produced in response to specific stimuli. Instead they may be part of the routing protocol of the system. Through the spontaneous activity of millions of neurons—passing a kind of keep-alive message, as described earlier—the default mode network could conceivably help the system work out accurate "routing tables," which can be invoked as needed to send actionable messages across the brain. Monitoring network status in this way requires constant updating with sparse messages, whether or not the brain is doing a particular task. Thus, it is possible that the default behavior is to periodically test communication links with neighbors, and to share information about more distant parts of the network, just as on the internet.

This kind of system would be an ideal substrate for learning based on the internet metaphor's framework. When we learn, we find a good route for messages to travel. But learning new routes requires an accurate assessment of the network's connectivity and status. This information needs to be gathered continuously, and in advance of working out a particular route. The default mode network could provide this knowledge, and do so at convenient times, like when the brain is not engaged in other demanding tasks. It's possible that sleep states could be another component of a system for monitoring network status and connectivity.

Finally, in our tour of flexible neural mechanisms available to the whole brain to perform routing, we have a solution that has been noted by many researchers, one that enjoys considerable acceptance already. It is the idea that synchrony in brain activity serves functional goals. This notion was first proposed in relation to the concept of "binding." The idea is that neurons in different areas trying to pass messages will generate firing that is synced in time between the two areas. This activity serves to "bind" different neural populations together. Much evidence has accumulated since the 1980s that this neural phenomenon is associated with functional outcomes.

For example, when we perceive a visual object, the neurons representing different parts of that object need to be bound together somehow. This binding allows us to see the object as a single thing rather than as a collection of parts. Working with cats, German neuroscientist Wolf Singer and American Charles Gray showed in 1989 that synchronized firing occurs when different populations of neurons in an area respond to the same visual quality of an object.[64] In particular, separate populations of neurons in the visual cortex sensitive to rectangles tilted at a particular angle will become synchronized when the preferred rectangles are presented.

But synchrony can do more than bind related information within a small chunk of the brain. As German neuroscientist Pascal Fries and others have noted, synchrony could also solve the problem of flexible routing among brain parts. Fries has been one of the earliest and most vocal advocates of routing in the brain. In a landmark 2005 paper, he wrote: "Now it is time to understand how the many active neuronal groups interact with each other and how their communication is flexibly modulated to bring about our cognitive dynamics."[65]

How could synchrony generate flexible routing? Across the whole cortex, we can imagine a patchwork of neural groups, and each group firing with its own rhythm. Routing by synchrony is the idea that populations whose rhythms are coherent have an excellent opportunity to exchange messages. As Fries argues, when there is a fixed interval between firing in one population and firing in another population, messages can be passed because activity at a certain rhythm in the sender makes it more likely that the receiver will be active in the same way.[66] "Activity" here could mean both spiking activity and also lower levels of electrical activity that oscillate in a population of neurons. In both cases, activity in one part of the brain predisposes neurons in another part to be active a short amount of time later (figure 6.7). This mechanism

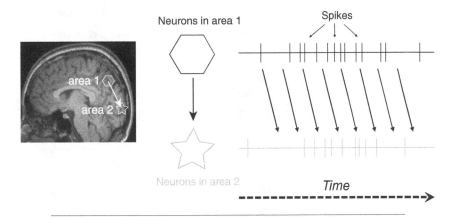

FIGURE 6.7 Routing by synchrony. Independent regions of the cortex can influence each other's pattern of activity by entraining in time. Spiking in one region (hexagon) leads to a similar pattern of firing in another region (star) some distance away. By selectively synchronizing in this way, the brain could exchange information dynamically without changing network structure.

Adapted from Pascal Fries, "A Mechanism for Cognitive Dynamics: Neuronal Communication Through Neuronal Coherence," *Trends in Cognitive Sciences* 9, no. 10 (2005): 475.

works both for populations of neurons that are directly connected to each other and also for ones that are more distantly separated. Synchrony between areas has been likened to a "carrier wave" that opens a flexible channel for communication.[67]

Somewhat counterintuitively, Fries argues that flexible routing is in large part about *preventing* synchronization. Stopping most areas from syncing allows the brain to select only the linkages needed for a particular task at a particular time. This is achievable because most neural oscillations persist for only a few seconds at most. In other words, routing by synchrony is a short-term, flexible strategy. Suppressing most synchronization also ensures that the whole system (or a big chunk of it) doesn't become synchronized. Excessive synchrony is what happens during an epileptic attack. This must be prevented. So despite bursts of activity in all parts of the brain from time to time, a fundamental goal of routing by synchrony is to make sure that most senders and receivers are *not* synchronized except when they need to be.

Routing by synchrony may seem incompatible with the internet metaphor since I have described the internet as asynchronous. However, though the overall system of the internet is asynchronous, its components can and do utilize synchrony. In order to pass messages between two routers along a path, they need to be in sync.

Indeed, routing by synchrony may solve a problem that has not been recognized: addressing. It seems unlikely that messages in the brain travel with the full network address of their destination, as packets do on the internet. Possibly, too little information is carried by spike trains to encode both a full address and "content," even if spike timing carries some of this information. So how does the brain find the right place to direct a message when it is traveling over more than one hop?

The answer may be that a particular rhythmic pattern of activity in a population of neurons can be seen as an address. It's possible that when a particular rhythm is produced, the goal of the brain's routing protocol is to find a matching rhythm somewhere else in the brain. This destination is automatically linked to the sender because of the way neurons are excited. It's as if one group of neurons is calling out to the whole network with an address. The "call" may be relayed over more than one synapse through the network backbone (the brainstem, thalamus, claustrum, and other core areas). The callee then lets the caller know where it is on the network by producing a matching rhythm. Once a route to a destination is found, messages can be passed asynchronously, rather than through strict synchronization as currently envisioned by the routing-by-synchrony model. New destinations can be chosen by producing the matching rhythm somewhere else. In this way, messages could flexibly travel over more than one hop and still be delivered reliably, quickly, and sparsely.

A decade and a half after Fries's landmark paper, neuroscientists may finally be at the verge of seriously investigating flexible routing in the brain. Theorists are increasingly demonstrating that the mechanism of routing by synchrony is plausible, and even of fundamental importance.[68] This approach can be taken much further within the framework of the internet metaphor.

• • •

To summarize, the flexible routing of messages in the brain could be built out of more than a dozen possible neural mechanisms. They include, but are not limited to:

1. Timing codes for the axon proposed by Scott
2. Axonal branching
3. Synaptic biochemistry
4. Glia

5. Nest/basket cells
6. Local gating circuits
7. Reticular formation
8. Thalamus
9. Claustrum
10. Frontal-parietal networks
11. Striatum
12. Drivers and modulators in the cortex
13. The default mode network
14. Routing by synchrony

Can all these potential mechanisms work together? Are these systems balkanized and specialized, or are they part of a grand system? Though they may appear incompatible with one another, we should consider that making sense of the signals in a black-box communication system is very difficult if we don't understand the basic protocol. At different levels of communication architecture, a message can take different physical forms. This is the case on the internet: a packet of data can be represented as a series of voltage pulses at the Physical Layer of the protocol stack, or as an array of capacitor charges (say, while it is waiting in a queue in the Transport Layer). The same message may take different forms in the brain as well. Understanding how this works is a challenge, but perhaps not an insurmountable one.

The small world–like network topology of the brain implies a unity of routing protocol, complex though it may be. Different brain systems—vision, motor, emotion, planning, language—are just too close to one another on the network to perform an array of highly specialized and dissimilar routing schemes. Though each system may be organized and connected in distinctive ways, they may all adhere to the same basic protocol. Recurring mechanisms of large-scale routing in the mammal brain—including the use of drivers and modulators, as well as routing by synchrony—also suggest a unified protocol. For example, the oscillations involved in routing by synchrony are found in many parts of the brain and operate on essentially the same timescale in all mammals tested to date.[69] Overall, these commonalities suggest a unified global strategy for flexibly controlling network traffic in the mammal brain.

• • •

Question 4: How do we add more message-passing nodes?

To conclude this chapter, we turn to the question of how to *scale* a brain network that performs flexible routing. Scaling occurs on an evolutionary level, so if we accept that flexible intercommunication is fundamental in brains and that a unified and efficient protocol is required, we are faced with a need to square this picture with the expansion of the brain in mammalian evolution. Scaling also occurs on a developmental level, as we will see. Thus, we need to ask how evolution and development can build brains that vary in size over time but that can still exchange messages widely and flexibly.

It is unlikely that basic routing protocol in the brain can be changed very fast through evolution, especially if the brain is using a unified protocol. By their very nature, large-scale communication systems are mostly stuck with the basic protocol already in use. Changing basic routing protocol is very cumbersome. For example, the circuit-switched telephone network was poorly suited to the packet-switched internet. Dial-up was a clever exploit, but it was unsustainable and cumbersome. Entirely new systems needed to be wired up across the whole network of communicators to build the modern internet.[70] For this reason, it is more likely that mammal brain evolution is based on elaborations of a single highly flexible and robust routing scheme, rather than on the emergence of many distinct protocols.

Mammal brains have gotten much bigger since the end of the dinosaur era 65 million years ago. From groups of mostly small shrew-like creatures with 100 million neurons or fewer, mammals have increased in size rapidly, leading to brains like the elephant's with more than 1,000 times more neurons.

The fact that mammal brains have dramatically scaled up over a relatively short time period means that we have a natural experiment that sheds light on the challenge of scaling. The question is, how do we scale up the mammal brain in a graceful way? In particular, how do we add neurons to the brain over evolutionary time in order to perform new functions, while preserving core functions shared with ancestors?

The mammal brain presents particular challenges because of its fundamental patterns of interconnectivity. It is unique in having a corpus callosum, which exchanges messages between the two hemispheres.[71] Mammal brains have also greatly increased the proportion of neurons whose axons are wrapped in myelin insulation (myelination), which helps to speed delivery of messages.[72] These may seem like advantages, but they also present a problem in bigger brains. The difficulty is that most neurons have lots of connections, so adding more neurons means adding lots more connections, which take up more and more space in the cranium for bigger and bigger animals.

The internet metaphor suggests that a unified—but fundamentally distributed—protocol is necessary for graceful scaling. While the internet does not have the same limitation on space that brains have, it has still shown the capability of efficiently adding more nodes. One reason is that the network topology of the internet and of mammal brains gives unified protocol a big advantage. Because it is a small world, adding neurons anywhere in the brain automatically provides access to the entire network over just a few hops. Once they join, new neurons could potentially receive network status information as well.

If messages are handled on mostly equal footing in the mammal brain, new neurons can perform new functions without having to build a new and possibly incompatible routing protocol. Though it may be unified, the brain's protocol allows for substantial local flexibility in implementing the protocol—just like the internet. For example, each local region could fine-tune its strategies for verifying message delivery (with acks or nacks as appropriate). And each region could offset the costs of transmission in such a way as to accomplish distinct goals.

As it turns out, the process of adding more neurons to the brain over evolution gives us a practical tool for testing whether the brain uses an internet-like routing protocol. Recall that routing schemes vary most in the degree to which they share information channels. A circuit-switched network like traditional telephony uses exclusive channels: when two communicators are on a call, no others can use the connecting channels. In contrast, a packet-switched network like the internet uses channels that are shared at all times and open to all comers. It turns out that a black-box communication system can be characterized in terms of this spectrum of sharing. Importantly, this can be done without knowing what the specific routing protocol in the black box is.

The trick is to see how the network behaves as it grows. As more nodes are added to a telephone network, with its exclusive links, no one gets slowed down. My telephone conversations aren't delayed if my new neighbor installs a telephone line to their house. On a traditional telephone network, there is no cost to me if the system adds more pairs of communicators anywhere on the network. New nodes can be added in a way that matches likely usage, and then everyone has a private channel.

This works well for a while, but at some point there are just too many connections. When this happens, lots of messages can't get through at all—we all get busy signals because all the intermediary links are in use. When the system is overloaded, successful message passers get their signals through just as fast as if they were the only ones on the network, but everyone else can't send messages at all. The system has reached overload. This problem follows from the fact that once a channel is engaged by a pair of communicators, it cannot be shared.

At the other end of the spectrum of routing schemes, we have the packet-switching protocol used on the internet. Here we have channels that are fully shared. My packets will inevitably mingle with those of my neighbors, and on equal footing. Because of this sharing, adding more communicators slows everyone down just a little. But as the network grows, the risk of total overload is low.

The different forms of scaling engendered by circuit switching and packet switching can be seen schematically in the top frames of figure 6.8. Here we are looking at a small, representative part of the network, not the whole small world–like network. When the network has fewer nodes, as in the first two rows, circuit switching gives new communicators a good chance of being able to seize a channel for their exclusive use. In packet switching, adding even one more pair of communicators using the same channel slows the others down (alternating shades in the middle bar indicate messages sharing the channel). However, when the network gets very large, as in the bottom row, circuit switching runs the risk of overload. New communicators are at risk of being shut out completely, or else of blocking those already on the network, as indicated by the many communicators who are unable to pass their messages (given by their shading). This is not the case for packet switching.

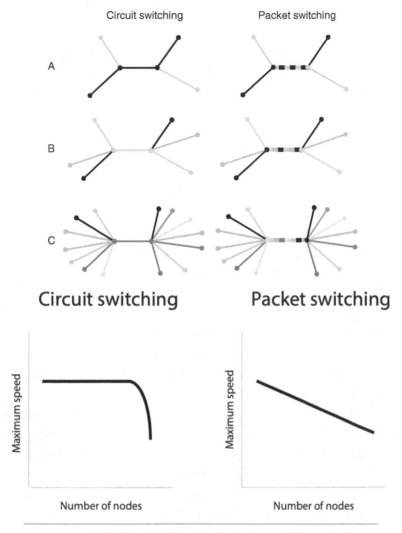

FIGURE 6.8 TOP: Growth of circuit-switched network like a telephone system (left column) versus growth of a packet-switched network like the internet (right column). BOTTOM: Speed of message passing as a function of network growth. In rows A and B, the circuit-switched network provides essentially the same maximum speed of transmission, since the central link is seized for exclusive use by the communicating nodes, and there are few enough communicators that congestion (busy signals) are unlikely. However, with many communicators trying to use the central link, as in row C, congestion is likely and the speed of transmission drops substantially. In contrast, a packet-switched network incurs incremental costs as more communicating nodes are added, resulting in a slowly decreasing speed of transmission. Image by Daniel Graham.

Engineers work very hard to avoid overload in communication systems. But evolution does not have the advantage of planning and forethought. Whatever routing scheme is utilized in the mammal brain has to contend with unplanned growth over evolution. While the mammal brain may not be in danger of total overload, its scaling could help us identify the global routing protocol in use.

If we were to graph how neural systems grow, we could compare the number of new nodes added and the maximum speed for a message to be passed (see bottom of figure 6.8).[73] By maximum speed I mean the highest rate at which a message can be passed from any brain part to any other brain part. For a circuit-switched brain, the line stays flat as we add more communicators. But at some upper limit there will be too many nodes, and it will quickly be hard for a node to pass a message. At this point, the line representing maximum speed will shoot down. In contrast, the maximum speed in a packet-switched brain will go down slowly and at a constant rate.

To test this in real brains, we need to make assumptions about both message passing and brain scaling. We will discuss the specifics of this question in the next chapter. But all else being equal, the maximum speed of transmission should have one of these two functional forms. It could be a flat function, possibly with a steep falloff, meaning there is little sharing of channels. This would indicate something like a circuit-switched protocol. I don't expect there to be a big falloff in extant mammal brains, but the danger of overload may be a constraint. Another possibility is that the function of speed versus size could be slowly decreasing, which would suggest extensive sharing of message-passing channels. This would imply a packet-switched protocol.

A practical test of brain scaling and network communication speed might show that the mammal brain is closer to an internet-like system than to a telephone system. As I have argued, a sparsely active brain is well positioned to take advantage of a packet-switched protocol. Most neurons in most parts of the brain can only be very active occasionally: short bursts are followed by long silences. This behavior further limits the chances of collisions and the potential of network-wide slowdown.

Since human brains are highly flexible and built on the same basic plan as other mammals, the flexibility afforded by packet-switched protocol would seem to be a precondition for our brains. But different mammal brains may also compromise between sharing and exclusivity. If neural messages are on the whole very small and sparse, channels may only need to be partially shared in smaller brains, while more sharing may be required in bigger brains like ours.

Whatever the correct answer may be, we would not have asked the question in the first place if we had stuck with the computer metaphor.

• • •

Finally, in addressing the question of brain scaling, we need to acknowledge that any solution to flexible routing must also comport with developmental restrictions.

During development, neurons and synapses are added, especially just before birth. But they are also taken away over subsequent months and years. At the beginning of development, when neurons grow and mature out of populations of precursor cells, they form dense connections with essentially every other neuron in their vicinity.[74] Some also extend connections to distant targets at around the same early stage.

Over time, many local connections are lost. The brain as a whole reaches an adult state with a characteristic pattern of connectivity. As we saw in chapter 4, across adult mammal brains there is a predictable relationship between the physical proximity of two brain regions and the likelihood and density of their interconnection. Nearby areas are very likely to be connected. There is a much lower chance that more distant areas are connected, but the strength of connection is fairly uniform over a wide range of distances. For the most distantly separated areas, the chances of connection drop rapidly.

How can a routing protocol support effective behavior as it develops toward this end state? We can extend the notion of learning in an internet-like network to the process of developmental scaling. As brains add and subtract neurons and connections, useful paths are activated, and messages can be passed on these routes as needed. The rich array of neural routing mechanisms I have described could help manage developmental scaling while also attaining learning and flexibility. There are also latent connections that can be called on during development when there is network damage, or when external conditions change. However, brain development—especially in humans—remains one of the most poorly understood aspects of neuroscience. As we learn more, it will be useful to consider routing protocol and internet-like solutions at different points in brain development.

• • •

The reorientation of viewpoint under the internet metaphor allows us to see the brain from the point of view of messages rather than neurons. If we do this, we have the new set of questions given at the start of this chapter: How are message collisions dealt with? How is reliable delivery of messages achieved? How is flexible routing of messages accomplished? and How do we add more message-passing nodes? I have tried to address these questions in terms of brain function as it is currently under-stood in order to build a framework for a more formal theory of mammalian brain function.

The next chapter will provide a critique of this framework. As we will see, one impediment to applying the internet metaphor is the distaste for theory in the biosciences, which traces back to founders of the field like Ramón y Cajal. Ironically, Ramón y Cajal, the devoted empiricist, provided one of the earliest examples of the success of theoretical extrapolation. We will also examine critiques of the internet metaphor framework from the perspective of computation, and we will return to thorny technical questions, such as what constitutes a "message" in the brain, how the brain performs addressing, and how we can leverage brain scaling to understand the black-box routing protocol of the brain.

7

CRITIQUE OF THE
INTERNET METAPHOR

Basically, the theorist is a lazy person masquerading as a diligent one.
He unconsciously obeys the law of minimum effort because it is easier
to fashion a theory than to discover a phenomenon.
—Santiago Ramón y Cajal, *Advice for a Young Investigator*

MANY criticisms of the internet metaphor are possible.
I will try to address a subset of these criticisms, beginning with a general defense of the idea that metaphor
and theory are useful in biology, in response to Ramón y Cajal's
quote. The rest of the chapter addresses specific criticisms. This
critique has been structured as a dialogue.

• • •

*You have argued that metaphor serves as a proto-theory and that
metaphor has been critical to progress in other sciences. But biology
can't be put in similar terms as physics. There are no laws that hold
always and everywhere. Even the most important and accepted
theory we have—Darwinian evolution—is endlessly debated in
its particulars, and there is no universally accepted formulation of
it. With so much still unknown about brains, shouldn't we just skip
theory altogether and focus on finding new phenomena in the brain
as Ramón y Cajal suggests?*

A theory is a collection of systematized knowledge that pertains to a given realm of the universe. Theory is abstract understanding. It is about some aspect of the system that we can't pick up and hold in our hand. A good theory encompasses and summarizes a large body of current understanding. In many sciences, there are clear delineations between theorists and experimentalists, and neither is expected to do the job of the other.

I have opened this chapter with the acid quote from Ramón y Cajal to highlight a long-standing distaste for theory (and by extension, for metaphorical frameworks) in some quarters of neuroscience. There is good reason for skepticism about theoretical approaches in neurobiology. Chance—on both the micro and macro scale—and nonuniformity are fundamental realities in all biological systems, including the brain. Because mechanistic theory is rather ineffectual in biology, we often revert to metaphorical thinking. Metaphors can serve a similar purpose as formal theory. But in the absence of theory we need to always search for better and better correspondence between biological systems and the metaphor we adopt. The internet metaphor is not itself a theory. Yet as a proto-theory, it prompts questions whose answers could form a coherent theory of brains.

The utility of theory in neuroscience is well articulated by renowned neurobiologist Eve Marder of Brandeis University. She has said that a good theory is not one that predicts the data best, but rather one that causes experimentalists to do a different experiment than the one they were planning to do.[1]

Ramón y Cajal himself, ironically, also provides an illustration of the usefulness of theory. He is justly considered a giant and founder of modern neuroscience, whose art (see figure 7.1) is nearly as spectacular as his contributions to cellular neuroanatomy.

Why was a scientist as great as Ramón y Cajal so opposed to theory? Ramón y Cajal opposed all forms of abstract theory,

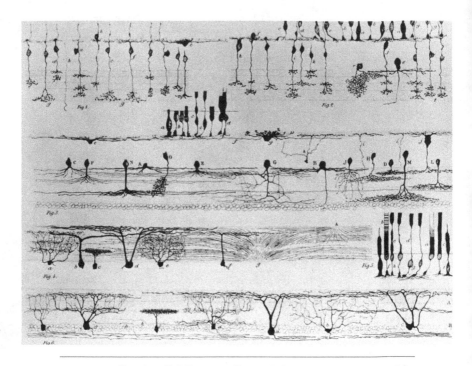

FIGURE 7.1 Ramón y Cajal's outstanding artistic output was on a par with his seminal scientific achievements. Ramón y Cajal, "Nerve Structure of the Retina," 1923, ink on paper. https://collections.nlm.nih.gov/catalog /nlm:nlmuid-101436424-img.

including theories of cosmology and fundamental physics. He argued that our brains are not adapted to natural philosophy, but rather to more mundane tasks of survival and reproduction, so we should not expect to ever solve these mysteries.[2] As the epigraph implies, Ramón y Cajal thought it was not even worth trying. The only meaningful thing to do is to intensely study a particular system and describe it as best as we can.

This view makes more sense when seen in the context of his time, the turn of the twentieth century, when empirical studies in anatomy, physiology, biochemistry, and ethology—and indeed in science more generally—were making enormous strides. In almost all areas beyond basic anatomy—from hormone chemicals to the basic divisions of the nervous system to brain diversity in different species—there was much to discover about basic processes. There was, as scientists say, lots of low-hanging fruit. Scientific innovation in biology was no doubt facilitated by rapid advances in technology and engineering.

But at the same time, scientific theory had become weird, especially in physics, where new theories of relativity and quantum mechanics challenged our basic understanding of the universe. Perhaps for this reason Ramón y Cajal saw theory as something to be avoided in biology, especially with so many new empirical phenomena waiting to be discovered. Moreover, it was certainly the case that metaphysical and unscientific approaches to biology and to what we would now call neuroscience were common in Ramón y Cajal's time, and based on little if any real evidence. Indeed, the history of brain science is replete with speculation. This is probably because the stakes are so high (what am I?) and relevant evidence (my experience and behavior) is so near at hand.

The irony is that one of Ramón y Cajal's most important contributions to neuroscience came from the kind of theoretical hunch he so disliked. Ramón y Cajal is celebrated today as the originator of the law of dynamic polarization. Not only are neurons individuated, he proposed, but also a given neuron can only send signals in one direction along an axon, usually from the dendrites and cell body toward the axon terminal. Since an axon is just a tube of liquid with charged ions in it, it could, according to electrodynamics, conduct signals via movement of ions in either direction (toward the dendrites or toward the axon terminals). But Ramón y Cajal argued that signals traveled only toward the axon terminals.

Ramón y Cajal had collected overwhelming evidence on the dense interconnectivity of neurons in many brain parts, in many species, and at different stages of development. But he reasoned that, if signals could travel either way along an axon, these kinds of networks would be prone to uncontrolled activity, with a single excitation leading to a chain reaction in many other neurons, and eventual overload. The problem would become intractable if signals could flow forward and backward in every neuron.

To limit the spread of activity, he proposed that signal traffic in neurons flows only in one direction, which is to say that neuronal signaling is polarized. He pointed to the structure of the retina and its pattern of connections to the visual brain as evidence for this one-way flow. But though it was possible in his day, he could not record neural activity in his laboratory. Instead, he formulated the law of dynamic polarization through theoretical inference based on anatomy alone, not based on observation of actual signal flow.[3]

Ramón y Cajal's conclusion required the assumption that there was some information for a neuron to transmit—not an unreasonable idea. But the notion of information processing requires abstraction and supposition. In other words, Ramón y Cajal's discovery came not from intensive empirical study of a system but rather from a theoretical elaboration. Though it doesn't require the fancy equipment and heroic patience needed to investigate specific phenomena in biological systems, abstract theory is nevertheless critical for scientific progress. Indeed, it is remarkable that Ramón y Cajal made such important contributions as an experimentalist and as a theorist, despite his protestations.

Today, almost all theoretical approaches to the brain are guided by the computer metaphor. As I have emphasized, the brain does perform myriad computations, and computational neuroscience is a crucial theoretical program. And there is no reason to stop looking for new phenomena. But without consideration of communication protocol, we are left without a broader theoretical understanding.

It may seem that I am making a contradictory argument by saying that neuroscience distrusts theory but yet is dominated by the computer metaphor. Both things can be true. We are skeptical of pure theory, mostly with good reason. We use metaphor instead to guide experiments. We ask: what computations underlie a given function, and what computations are performed in a given neuron? This approach has been successful. But by distrusting theory, we are blind to how much we are guided by sensible but mostly unscrutinized metaphorical frameworks. The problem is that we haven't confronted the limitations of the dominant metaphor yet by offering a new one.

• • •

In the last chapter, you made several comparisons between strategies for routing messages on the internet and ways these strategies could be implemented in the brain. But these proposals aren't very specific, or else they require data that we can't feasibly collect today, such as tracing signals as they traverse brain networks. Is there any neurobiological evidence to support your proposals today?

There is already research that points toward internet-like routing in the brain. Despite the technical challenges, innovative researchers have devised ways to trace the flow of messages in brain networks. For example, an experiment by neurobiologist György Buzsáki and colleagues using rats was carried out in 1999.[4] These researchers were able to follow spike sequences within a local networks of a few neurons. The spike sequences, consisting of a handful of spikes in a row separated by a distinct pattern of silences between spikes, popped up at different times in different neurons. Remarkably, Buzsáki and colleagues were able to follow these messages as they traveled among neurons. These traces of activity (Buzsáki has more recently called them "ripples"[5]) are a rarely achieved but crucial piece of evidence for understanding the brain.

Buzsáki and colleagues were able to watch network communication because they made several concurrent recordings of the same handful of neurons. This is like listening to the many conversations at a party with several microphones positioned at different points in the room.[6] In this situation, all the voices are overlaid. Some people will be closer to a particular microphone, and their voices will be louder, but they will also show up more quietly in microphones further away. With enough microphones, each speaker can be triangulated and his or her speech isolated. Then a message passed from one person to another—like in the game of "telephone"—can be followed as it travels.

Buzsáki and colleagues did the same thing with electrodes in the vicinity of the target neurons. Each electrode listened to voltage changes of the neurons nearby. To trace a message of a few spikes, the researchers made some assumptions about how activity in the different neurons was related. They scrambled the spike recordings in a variety of ways to determine the most likely pattern of message passing among the neurons. This intensive computation took the equivalent of twelve straight days of computer processing using the fastest workstations available at the time. While the researchers could not trace the physical connections among the neurons, their analysis gives good reason to believe that the neurons they listened to were sending messages to one another. Buzsáki and colleagues could measure with high time resolution where a given message was at a given time.

The researchers were recording in the hippocampus, an area whose main job seems to be keeping track of where we are in a particular environment. It is thought to tie specific locations in space to rewards, dangers, and actions in those locations. In Buzsáki's study, some patterns of message flow between the neurons seemed to be associated with times when the rat was running in a wheel for a reward, but only when the wheel was placed in a particular location in the rat's environment (figure 7.2). Distinct patterns were also found during sleep.

From this set of data, one thing should be clear immediately: the pattern of message passing among neurons is a major component

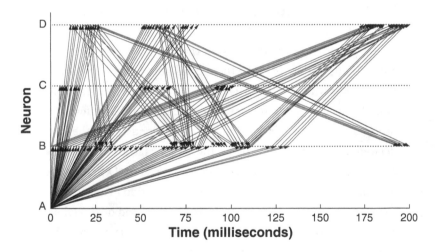

FIGURE 7.2 Exchange of messages among four neurons over time in the rat hippocampus. Messages pass among four neurons, labeled A–D. Each message's transmission is indicated by a line. Horizontal axis indicates time (0–200 milliseconds). Note that messages can take different routes to the same destination, or take travel to different destinations, and the time delay at each hop can vary considerably.

Redrawn with permission from Zoltán Nádasdy et al., "Replay and Time Compression of Recurring Spike Sequences in the Hippocampus," *Journal of Neuroscience* 19, no. 21 (1999): 9501.

of how the brain does its job. If the researchers had recorded only the sheer number of spikes produced by the neurons, they would have missed the rich network communication going on.

We can go further in inferring the brain's communication protocol based on these data. In figure 7.2, there are four neurons represented on the vertical axis, labeled A, B, C and D. The lines represent individual messages traveling between the four neurons. The plot is a collation of data from several recordings.

Notice the many messages that travel from neuron A to neuron D, and then to neuron B. First, we have different messages passed on the same route with varying delays. Numerous separate messages are passed from A to D, arriving within a few milliseconds. But others take longer to get to neuron D before traveling on to neuron B. The variation in delay for the first hop is considerable.

Messages with similar delays on the first hop can take different amounts of time to make the second hop. Several lines travel from D to B with a similar delay. At neuron D, some messages are passed to neuron B right away, some are delayed just a few milliseconds, some are delayed about 50 milliseconds, and others are delayed around 175 milliseconds. The same behavior on the same route can be seen in later-arriving signals at neuron D that also go to neuron B. Here again messages show a variation in delays, with some messages going from D to B very quickly and others taking considerably longer to make it. There are also alternate routes that lead to the same destination. For example, messages can travel from A to C to B and directly from A to B. While all this is happening, it remains possible for neuron A to send messages to a different destination two hops away: A to B to C or A to B to D.

To summarize, Buzsáki's experiment suggests that a single neuron can do all of the following:

- Send messages on a given route with a small variation in delay
- Send messages on a given route with a large variation in delay

- Send messages on different routes to the same destination
- Send messages to different destinations
- All of the above on multi-hop routes

Clearly, a single neuron has a lot of flexibility in its message passing.[7] To behave in such a flexible way, it could use one or more of the mechanisms discussed in the last chapter, such as axonal branching dynamics, glia, and multineuron gates. In Buzsáki's data, the spacing of messages in time is consistent with a back-off algorithm like those of the internet. Consider the small variation in delay for messages traveling on the same route. Some of this timing jitter might serve to avoid collisions with other messages.

However, some messages also displayed large variations in delays along a route. This behavior is consistent with the idea that neurons have the ability to store messages for a time—a kind of buffer. Buffering is a fundamental requirement on the internet. For example, back-off algorithms dictate that routers have the ability to store messages for a brief time: messages that collide must still exist at the sender so they can be resent.

While buffers have been seen as unrealistic in brains, they would confer great advantages in terms of robustness. As on the internet, buffers in the brain could be small. They need not store much information. Buffers also need not store messages for long. On the internet, packets that get "stuck" in a buffer are automatically deleted after a fixed amount of time.

In the neurons Buzsáki and colleagues studied, buffering would only require storing a small amount of data for a few dozen milliseconds. Since the connections among the studied neurons in Buzsáki's experiment were not mapped, it is unclear whether the substantial delay in passing a message is due to the sending neuron buffering the message itself before sending it directly to its destination, or due to the message being passed through one or more intermediaries. I believe the latter case is more likely because neural mechanisms that could support flexible routing may also support buffering, perhaps by passing buffered messages in a precisely timed local loop.[8] In any case, either or both mechanisms could accomplish buffering.

Finally, both the network studied by Buzsáki and the internet exhibit occasional spontaneous activity. The neurons produce one or two spikes every once in a while. Because they are not part of a repeating sequence of spikes, the researchers were not able to trace the propagation of these solitary spikes. But it is possible that these signals serve a purpose akin to acks or keep-alive messages on the internet. Given the network behavior Buzsáki observed, acknowledgment and network status seem possible given the multiple delays, paths, and destinations a single neuron is capable of utilizing. Buzsáki's data do not prove that internet-like communication protocol is in use in the mammal brain, but they are consistent with primary strategies used by the internet.

• • •

You discuss single neurons but have mostly ignored brain imaging and other whole-brain experimental measures like EEG (electro-encephalography). Moreover, many scientists study the architecture of whole brains. Are all neuroscientists really as fixated on the single neuron and its computations as you say?

There have been numerous avenues of research aimed at understanding the whole brain. These have included both anatomical studies and studies of brain activity.

Intensive anatomical investigations of gross brain structure go back hundreds of years, and less formal observations go back thousands of years to ancient Egypt. In recent times, comparative anatomical approaches have looked at the evolution of brains and their development in many species. From this work, we have learned that brains follow certain basic organizational principles. For example, there is a fixed relationship between the overall size of the mammal brain and the size of just about any brain part.[9] As a consequence, the main way for evolution to generate a larger cortex or a larger cerebellum is to make the whole brain bigger. Evolutionary relationships of this kind are important in internet metaphor–guided neuroscience.

We need to understand activity in living brains too. Yet most physiological approaches to studying the brain do indeed see it as an agglomeration of single neurons performing computations, which are further organized into parcellations that perform a particular computation. This theoretical bias stems from the computer metaphor. It is compounded by major limitations in measuring whole-brain activity.

There have been several generations of noninvasive electrophysiology like EEG, as well as increasingly powerful brain imaging. But imaging methods like fMRI have fundamental limitations that largely have to do with their resolution in time and space. Imaging cannot resolve details on the level of single neurons or local neural assemblies. We can return to the analogy of the time-traveling engineer we met in the preface to this book. If studying a part of the brain is like figuring out what a router circuit board does, brain imaging would be like using a thermometer to do this.

The limitations of brain imaging are not the primary focus of this book, but the limitations relevant to intrabrain communication are worthy of some discussion. Functional imaging measures blood flow over time, which is seen as a proxy for spikes, since more spiking requires more energy and chemicals from the blood. Blood flow in each small chunk of the brain is assumed to compute something: how much of a particular emotion the subject is feeling, how much an image resembles a face, and so on.

In a typical fMRI study, each datum of inferred neural activity summarizes the firing of hundreds of thousands or millions of neurons (albeit imperfectly: inferred neural activity is dependent on numerous assumptions of varying validity about the linkage between blood flow and spiking). In EEG, each electrode summarizes signals from many millions if not billions of neurons.

While each neuron operates on millisecond time scales, fMRI data are one thousand times coarser. Whatever it is that constitutes a message in the brain will probably be invisible to imaging. Nevertheless, imaging researchers have begun to consider how different parts of the brain communicate with one another. If two chunks are active in a similar way over time, they may well be communicating with one another. This is called functional connectivity. But measuring which direction the information is flowing between chunks is not possible with imaging: one cannot tell which was the sender and which the receiver, again because of low measurement resolution in time.

One might expect that all this imprecision in measuring neural activity would at least lead to a consistent, albeit blurry, picture. Unfortunately, fMRI generates quite inconsistent data even if the same task is repeated by the same subject. Say we measured blood flow in a subject while they recalled a list of memorized words. If we repeated the same experiment two weeks later with the same subject, we would be very lucky if measured blood flow were correlated at a level of 0.7 with our earlier measurement (1.0 being a perfect correlation). Even apart from the blurriness, this suggests that a lot of what we measure in an imagining experiment has little to do with the way brains actually deal with information.

With its emphasis on quantities of signals, the computer metaphor implies that fMRI data should be expressed and understood in terms of "more" and "less" activity in chunks of neurons. When there is more activity in a chunk, it is assumed that the chunk is more involved in the brain computation under study (reading, emotion, memory, perception, and so on). But why must this be true? There is plenty of evidence that regions of the brain like the visual cortex are less, rather than more, active when accomplishing a task, as was shown in 2008 by vision scientists Fang Fang, Dan Kersten, and Scott Murray using fMRI. Fang and colleagues found that activity in the primary visual cortex goes down when we perceive parts of a moving object as constituting a single "thing" (this is the binding of object elements mentioned in the previous chapter).[10] Lower activity is also observed in the default mode network when people improve performance on a task or when they encode a memory.[11]

In other words, when the brain does its job successfully, it can become less active in key areas. This finding doesn't fit well with the computer metaphor but is fully in accord with the internet metaphor. Whether there is "more" activity at a node when it is doing its job is not the most important question. What matters instead is how messages are directed across the network. For example, when we see "more" activity (more spikes, higher imaging response), it may be related to network maintenance (testing and sharing routes) rather than to the delivery of particular messages.

The resolution problem in imaging also affects measurements of brain network structure that exploit the same equipment. In structural imaging approaches such as diffusion imaging, the goal is to trace axon tracts connecting brain regions. Imaging is especially attractive as a way of mapping the connectome of humans, where injecting tracers in living people is not permissible.[12] Pairing of functional and structural imaging has also become more routine in recent years and allows us to study the architecture and gross activity of neuronal networks in a single human subject. The Human Connectome Project in particular has produced an abundance of such paired data. We will look in more detail at this approach in the next chapter.

But standard imaging techniques that trace axon tracts also have strong limitations. For example, they generate as many as 50 percent false positives; that is, half of the connections detected by diffusion imaging may not exist.[13] This leads to unreliable connectomes. Standard structural imaging techniques are also mostly blind to axons when they are intertwined, which is common in the brain.[14] In addition, structural imaging measures undirected connections, so once again we don't know in which direction information is flowing between two areas. This uncertainty can lead to further inaccuracies in models of the connectome.[15]

Even if MRI technology improves (which seems unlikely since the magnetic field strengths required approach engineering limits and become hazardous for living creatures), we still need to augment our theoretical approach. We need to move beyond measuring only where a computation takes place, as evidenced by an increase in activity. Instead, we should think about how brain activity reflects the brain's overall strategies for routing messages.

• • •

In chapter 5, you demurred from offering a clear definition of what you mean by "message." So what exactly is a "message" in the brain?

There is a broad range of candidates for "messages" at many scales, from single-neuron chemical interactions to whole-brain electrical oscillations. Routing systems that control the flow of all these kinds of messages must work together, but in ways that can't currently be described in formal ways. However, I think we can identify something that all of these types of "messages" have in common. I propose that a message in the brain can be defined as contiguous, directed activity linking several neurons in sequence over short time scales.

The particular activity pattern exhibited by each neuron in the sequence—its message "content"—could be complex, and it could be different at different points on the path (or paths). So at present it may be more tractable to limit ourselves to the path(s) a message takes and to see this as a proxy for the message itself. Paths are not necessarily between a solitary sender and a solitary receiver. They can have many branches and involve many neurons, as well as other cells like astrocytes, along the way. On the internet, what matters is, in large part, how messages are passed and what routing decisions are made along the path to direct message content in a particular way. Message content, on the other hand, may be more related to where a message originated or what it will do when it gets to its destination, just as message content on the internet is only relevant to end-user applications. Message content will also surely reflect the computations occurring at individual neurons (which are no doubt important).

For example, consider a neuron in the visual cortex. This cell doesn't know what the pattern of light is in the world. It only knows that a collection of photoreceptors, ganglion cells, and thalamic cells have passed it a message. If we record from the cortical neuron, the responses might look like a computation of the photonic structure of the visual world. This may be a useful transformation of information from the world. But the cortical neuron doesn't care about what the information is—how could it? What matters to the neuron is where the information comes from and where it should go next.

The goal of neural decoding under the computer metaphor is to understand what a signal means. At least in the visual system, this approach has achieved some success. But one advantage of using the internet metaphor and seeing the brain as a communication system is that it is not necessary to understand the content of a message. What is important is how messages are passed across the network.

• • •

So are neurons like routers, or are they like applications or users?

Neurons acting as senders and receivers probably have the dominant effect on single-neuron activity. But neurons may serve as routers as well. Some do only one of these jobs, while others may serve more than one of these functions at different times.

On the internet, there are far more senders and receivers than routers. In the brain, there may need to be more routers given space constraints. This accords with Murray Sherman's picture of the brain as possessing "drivers and modulators." Modulators are seen to control message passing—and are far more numerous than the senders of messages (drivers). In other words, most neurons seem to serve an intermediary function of routing messages to and from "users" or applications such as sensation, memory, and motor systems.

• • •

How might the brain perform addressing?

There are two basic ways to address a message. One is to do addressing separately from sending the message. This is what we do when we make a phone call. First, we secure the line by dialing the number of the person we want to reach. Once the line is seized, addressing is complete, and all subsequent communications travel on this route automatically. But there is no sense in which the communications themselves know where they are going, or could find the right path if they got lost. My voice, encoded in voltage changes, is transmitted faithfully as long as the link is established, but only to the phone I have called. There is no chance for my message to go to somewhere else.

The other approach is to tag messages with where they are going. For this to work, intermediary nodes need to inspect messages and direct them appropriately. This is what happens in postal systems. An entire message, like a letter, is wrapped in an enclosure with the exact destination emblazoned on it. This approach allows the system to handle many messages over the same "line": mail trucks and sorters direct large batches of signals at once. Addressing in this way generates delays since a message can sit in a queue at the post office for a while. However, it won't lose its address. Nevertheless, a related drawback of this system is that routes tend to be set early in a message's journey and are applied to a truckful of messages.

The internet uses a similar approach, but instead of addressing an entire message, it addresses each small part of a message or packet. Routes can also be changed much more easily than postal systems. This allows more flexibility and speed.

In the brain, addressing may be a compromise between a telephone-like system and an internet-like system. It's possible that routing by synchrony could serve as a kind of "call": when one area adopts a particular rhythm, it is in a sense dialing the number of a destination. But synchronization in the brain doesn't last very long—a few seconds at most. The messages themselves—bunches of spikes—are small and sparse, like packets on the internet. So a routing system that can handle sparse messages during short calls could be the best compromise.

Addressing can also exploit hierarchical organization. Network scientists Peter Dodds, Duncan Watts, and Charles Sabel have proposed the idea of *pseudoglobal knowledge* as a way for nodes in a hierarchical network to address the entire network without needing to know exact addresses.[16] In a hierarchy with inferiors and superiors, there is a trade-off between our rank and the specificity of our knowledge of subordinates. Those on the bottom know exactly how to get to the top via each of their "bosses." But the higher we are, the more subordinates will be part of our subnetwork. We are less likely to know exactly where they are on the network, since they will be several (downward) steps away. In the brain, which also possesses hierarchies, it is possible that messages are sent toward a specific target, but without a specified route. It's also possible that messages could carry the destination address, perhaps encoded in spike timing. A sender would then only need to direct a message to the right subnetwork, and nodes further down the hierarchy would sort out where exactly it needs to go. The internet does the same thing. For example, if we send an email to a receiver behind a firewall, routers outside the firewall will not know the exact address of the receiver. Instead, the route will be planned up to the firewall boundary, and the subnetwork will then route it to its final destination.

A final possibility is that there is no addressing in the brain. Perhaps the passing of messages is so well organized—through evolution, development, and learning—that every activity pattern is unique and reliable, finds its way without assistance, and never interferes with any other pattern. This is within the realm of possibility but manifestly unlikely. Thus, a conclusion we can draw from the internet metaphor is that in the absence of a robust and flexible communication protocol—with systems to avoid collisions, correct errors, sense network status, find paths, and so on—the brain must be an ideal signaling system, one that is almost flawless and incapable of error.

●　●　●

How could one test your proposal in chapter 6 that evolution-ary scaling can be used to investigate the black-box protocol of the brain?

The idea here is that basic routing protocols have different scaling properties. At one end of the spectrum, we have circuit-switched protocol, like a traditional telephone system. In theory, such systems scale with no cost. Network traffic is not affected as more pairs of communicating nodes are added. This works until the point where there is an overload and no more new pairs of nodes can be added. At the other end of the spectrum, packet-switched networks like the internet have incremental costs for network growth. Since everyone's packets are interspersed as they travel, each new pair of communicators imposes a small cost on everyone else on the network. But there is no precipice the net-work is approaching as it gets bigger (see figure 6.8).

Thus, communication networks scale in different ways depending on their routing strategies. What effect does scaling have on the brain's routing strategy? We can't add neurons to a living brain, but evolution has provided a natural experiment on the scaling-up of brain networks. Treating the communication system of the mammal brain as a black box, we might be able to deduce the kind of routing scheme in use by measuring changes in signal traffic across evolutionary scaling.

This idea requires some unpacking. Over evolution, the mam-mal brain has expanded from millions of neurons to hundreds of billions. Let's assume that a primary goal of mammal brains large and small is to be able to pass messages from just about any node to just about any other node. To accomplish this, we can suppose that all mammal brains use the same protocol, and that the proto-col falls somewhere along the spectrum of basic routing strategies employed in engineered systems. We can make predictions about how network traffic would change as we add more nodes.

Then the question becomes what exactly we would measure. In chapter 6, I suggested that a reduction in maximum transmission speed could be related to scaling. However, it is not clear how to extract this quantity from the brain. Instead, we may be able to use the sparseness of different brains as a proxy for network traffic, and therefore speed.[17] This means recording from thousands or millions of neurons at the same time in a living animal. Essentially, we want to know what fraction of cells are active at a given time.

Because recording from many neurons is very difficult, little is currently known about the precise level of sparseness in mammal brains.[18] For one thing, sparseness probably varies within a given brain and at different times. But collecting these kinds of data is not outside the realm of possibility. We will see in the next chapter that bioengineers believe it is possible to record electrical activity from every neuron in a living mammal with existing technology. By doing this in many species, we could see how sparseness changes in corresponding regions of different brains. If all mammal brains use a common protocol, we should observe changes in sparseness as brains get bigger. The pattern of changes could tell us about the mammal brain's routing strategy.

Consider first a circuit-switching strategy (similar to telephony). For a given set of "live" messages, the system would seize all the links necessary for passing each message, presumably without a lot of busy signals. All else being equal, this means that as we add more neurons, the total number of messages being passed at a given time should grow at the same rate as the number of neurons. Under circuit switching, then, the number of messages being passed at a given time should be proportional to the number of neurons. Assume that highly active neurons pass messages at their maximum speed. Assume further that the number of neurons is proportional to the number of highly active neurons.

Then we can estimate the effect of brain scaling on sparseness. As brains get bigger, from mice to elephants, sparseness should grow at the same rate as brain size. That is, the same fraction of neurons will be active in small and large brains, perhaps about 10 percent. If lots of messages collide in the biggest brains, we may also see a flattening or decrease in sparseness at this scale.

At the other end of the spectrum of routing strategies is packet switching. In this case, adding more nodes means mixing together more and more message packets from different senders. Because these packets would be distributed to make use of the network's architecture, they would recruit a large and increasing number of neural connections. Therefore, increasing proportions of neurons would be highly active at a given time. Sparseness would thus go down as brains got bigger. In other words, a larger and larger fraction of neurons should be active at a given time.

However, it is also possible that the mammal brain compromises between circuit switching and packet switching. In some sense, the main question is: what is the time scale of a "call" (i.e., a tranche of messages) relative to the time scale of a spike? If calls are usually very short, a circuit-switched system starts to resemble a packet-switched system. As I have suggested, a message in the brain could consist of a few spikes or bursts of spikes. If this were the case, the brain could have the best of both worlds: exclusive links during a call, which makes routing easier, and sparse behavior, which allows for greater robustness. In this case, we may observe only a slight decrease in sparseness as brains get bigger.

These predictions remain vague, in part because a more formal and specific prediction would require that we consider many factors that I have ignored. Some of these factors relate to the relationships between brain size and other quantities such as neuron numbers, neuron size, neuron density, synapse numbers, network topology, and other factors. These relationships are complex, and relevant data from mammal brains are limited. Each of these scaling relationships could affect the prediction of how network traffic would scale under different protocols. Then there are many other quantities whose scaling is not understood at all. Quantities like the number of messages per neuron, number of message *copies* per neuron, typical path length, and so on are not even defined currently, let alone measured.

Future theoretical and experimental work guided by the internet metaphor may help elucidate these relationships and allow us to make more concrete predictions. As we will see in the next chapter, new technologies like optogenetics may also allow us to scale up the activity in brain networks, rather than the size of the network. Scaling up the number of active neurons in a given network could achieve the same kind of test as adding more neurons.

For the moment, it will have to suffice to predict that, if the mammal brain uses an internet-like protocol, network costs will slowly increase as it scales up. This increase may be reflected in the sparseness of brain activity in different species. The purpose of a framework like the internet metaphor is not to make specific predictions. Indeed, what specific predictions are made by the computer metaphor—or the clock metaphor or the plumbing metaphor or the mill metaphor? Although prediction is an indispensable part of scientific theory, metaphors are useful insofar as they offer new and useful points of comparison between systems. These correspondences can in turn be applied and formalized to generate concrete predictions.

· · ·

Computers also need to transmit information quickly, flexibly, and reliably from one place to another and to many destinations.

How does intracomputer communication compare to intrabrain communication?

Computers are ostensibly concerned with computing. As such, I have argued that classical computers are not a good reference point for the communication requirements of brains. But modern computers do need to solve communication problems.

We saw in chapter 5 that computers use buses to move information around (see figure 7.3). Buses are shared wires that move many numbers back and forth. If there are only a few possible senders and a few possible receivers, a bus is a simple and effective solution. Any of several nodes can load data onto the bus, and the link is then seized until all the data arrive. Senders (e.g., processor chips) and receivers (e.g., memory chips) are each highly interconnected within themselves, comprising millions or billions

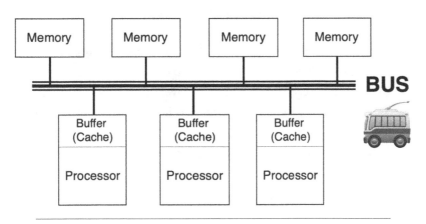

FIGURE 7.3 A computer bus that connects processors and memory chips. Each chip can load signals onto the bus for delivery, but only one batch at a time. A central controller (not shown) regulates the transmission of signals. Because requested signals may need to wait before transmission, there are memory caches in processors that store signals for a relatively short period before they get on the bus. Image by Daniel Graham.

of electronic components. The bus need only deliver a batch of information to the right chip, and then it can easily be distributed to the correct memory allocations or processing subcircuits.

Since reliable transport of information on the bus is so important, one might wonder how buses ensure delivery. As they are used in computers today, buses are so effective that there is no need for acks to verify that information has arrived at its destination. Consider what happens when we want to save a file, such as a document we are writing. We ask a processor to send some data to memory chips over a bus. For these kinds of messages, there *is* actually an acknowledgment that gets recorded. The acknowledgment will confirm that a chunk of data has been put on the bus. But this acknowledgment—called a "posted write"—doesn't actually confirm that the data made it to memory. In fact, posted writes are recorded at the same time the data are put on the bus, which is to say before they get to the destination.

Because the traffic laws related to buses are very strict, collisions are almost impossible, and return messages like true acks would be extraneous and inefficient. The system assumes by default that all message transfers on the bus from processor to memory are successful. When data travel in the other direction, from memory to the processor, there is no need for internet-like acks. To get data from memory, the processor first has to send a message to memory requesting the data. If the requested data arrive at the processor, this itself serves as confirmation that the initial request message was delivered.

Could the brain use buses? It's possible but unlikely. First, groups of riders on computer buses can come from several possible sources. The files they constitute can contain words, images, encoded instructions, or other data. Files getting on the bus can also come in any size and must travel together. Computer systems are well suited to these conditions. But the brain is not. There is nothing in the brain that resembles a single trunk channel that can (1) handle many different kinds and sizes of signals, (2) send and deliver them in two-way fashion to several destinations, and (3) operate almost continuously. If used in the brain, bus-like systems would also be overwhelmed by random physiological fluctuations. In computers, the rules buses obey eliminate randomness to such an extent that successful delivery is assumed.

Buses also impose inflexible rules on the chips they interconnect. While data are on the bus, no one else can use the bus. This typically means that data must be stored in buffers, which can be imagined as people waiting at many stations along the bus route. Buffers store all the data waiting to be transmitted. Fortunately, computers are not crippled by all this waiting because passengers get on and off the bus very quickly. The clock speed of computers is high, and wires transmit data at close to the speed of light. But still computer buses have large buffers available (called caches). The brain lacks all of these advantages.

If buffer-like mechanisms exist in the brain, they are likely to be small (though many mainstream neuroscientists would doubt even their existence). Information could be stored in membrane potential (the difference between the charge inside and outside the cell) or in local looping circuits, maybe for a fraction of a second, as described earlier in this chapter. A few neuroscientists have also proposed something similar in working memory systems.[19] I proposed in chapter 6 that neurons in the reticular nucleus of the thalamus are well positioned to perform this function. But mechanisms for storing anything more than a few spikes-worth of information, potentially for seconds or more, are less plausible. Mechanisms proposed for long-term memory in the brain, such as modifications of synapse strengths (Hebbian learning), operate too slowly for this kind of storage.

In contrast to the buffers available to buses, buffers on the internet are small, and the routing rules are less strict. Routers can store a relative handful of packets at a time in a buffer in order to manage collisions. But if packets stay too long or if they encounter buffers that are already full, packets can simply be deleted. In this case, the sender will know to send the lost packets again because it won't have received an ack. These kinds of behaviors seem more plausible in a system like the brain that works on much slower time scales than a computer, and one that likely loses a substantial number of messages. We will return to the concept of buffers in the next chapter.

In addition to buses, computers have historically used something called a crossbar switch to transmit information (figure 7.4). In this system, dozens of wires are laid in a square grid so that each wire crosses every other. At each intersection, there is a switch that allows the signal in one wire to be transferred to the wires it

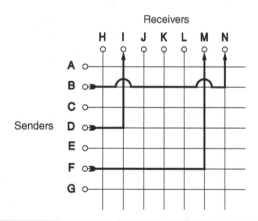

FIGURE 7.4 Crossbar switch. Senders of signals (A–G) can be selectively engaged to the array of receivers (H–N). However, only one sender can connect to one receiver. As with a bus, a central controller (not shown) regulates the connections between senders and receivers. Image by Daniel Graham.

crosses. By default, every switch is not engaged. When one or more switches are engaged, a direct route is established, say, between memory and processor.

The decision about which input is engaged with which output requires an external agent, called a controller, that makes all the decisions about which switches are open and closed. Crossbars require significant buffers because channels are used exclusively. When a message travels from one node to another, no one else can use the horizontal or vertical wires on which the message travels.

Because all those switches are costly to manufacture, crossbars are not often used in PCs today. But crossbars and similar components have long been used in telephone systems as switchboards and remain essential components of communication systems. Although they are a simple solution, however, crossbars are even more implausible in the brain than a single bus. Like buses, crossbars require substantial buffers and guaranteed delivery, but they are also dependent on central control. The decision to close any of the switches in the grid must be made by a single agent in real time (buses also require central control, but they typically manage a smaller number of communicating components). Neurons in some brain areas like the hippocampus do resemble a crossbar architecture. But this seems to be a superficial similarity rather than an indication of crossbar operation.

Thus, while computers must solve communication problems, they do so in fundamentally different ways than the internet. Computer architectures do not tolerate data loss, have substantial buffers, and perform switching centrally. These mechanisms are less plausible in brains than the solutions adopted by the internet.

• • •

Deep learning systems are computing machines that can become arbitrarily good at characterizing just about any set of data. Why not wait for them to solve the problem of how brains work? Someday, when we have ultramassive datasets, deep nets will give us near-perfect simulations that can predict the computations occurring in real brains. Humans may never understand the way deep nets make these predictions, but the fact that they succeed will be proof that the computer metaphor-based model is correct. Only if computers can be proven to fail at this task do we need to look for new metaphors.

Deep nets are useful to humans in many ways, and will become more so. Some scientists hope that we can "solve the brain" by collecting massive amounts of data using traditional methods (such as recording from neurons while a relevant stimulus is presented, or using brain imaging) and asking a deep net to model these data.

To some extent, big data approaches have worked before and might be expected to work again. Genomics, which is performed on the fastest computers and increasingly makes use of deep nets, provides the inspiration and the suffix for connectomics. But before scientists invested in the computational machinery to map the genome, they already had a correct guiding theory for how genes work, namely, that they are made up of base pairs within DNA that code for amino acids. We don't have the same thing for brains. We are still debating fundamental questions such as what a spike is for or how a brain region is defined. In any case, brute force computation in relation to genes has led to far fewer breakthroughs than were anticipated. The expectation that unique base-pair sequences would be associated with particular diseases or outcomes has mostly been disproven.

The resources necessary for approaching the brain in this brute force way are much greater. It would be risky—what if this doesn't work? The earth probably can't support the carbon costs of a lot of massive deep net–powered simulations of the brain. The brain would require huge numbers of highly sophisticated deep nets for even a basic model. Then it would have to be trained. Today, training a deep net that performs a limited task like translating a sentence from one language to another uses as much carbon as the manufacture and lifetime use of five fossil fuel–powered cars.[20]

Even if we could solve the brain with a deep net, we would not necessarily understand the brain any better. Since a deep net "model" is etched into huge numbers of weights—perhaps trillions of them if we were modeling the whole brain—how would we summarize and generalize this knowledge? As neuroscientist and AI researcher Vivienne Ming has written, "it is ludicrous to assume these algorithms will solve problems that we do not understand."[21]

This doesn't mean we should stop making computer metaphor–guided models using bigger datasets. This approach can be useful. But other theoretical paradigms should be tried out, too. We need better (but still imperfect) metaphors—and eventually new theories—to guide our investigations before we can fully take advantage of deep learning approaches.

• • •

A crucial aspect of brain network organization is the cortical column—the assembly of neurons that is tiled like a honeycomb across much of the cortex. Each column has much the same structure and resembles a module. There is a lot of interest in columns in part because of the Blue Brain Project, which aims to model every element of a single column and simulate its activity in great detail. For one thing, cortical columns suggest that brain networks are modular, like a computer. How do cortical columns square with the internet metaphor?

Cortical columns are worthy of explanation. They are interesting in part because they are mostly invariant across mammal species. In mice and monkeys, neurons form columns of around 100 to 200 neurons each. The neurons in a column are primarily connected to each other. Columns are also organized into common layering schemes that are repeated across the cortex. Some layers are for input and others for output, often to particular target regions.

But there have long been doubts that the column has a large functional role in brains. For example, in one monkey species, individuals run the gamut from almost no columnar organization in the visual cortex to fully ordered columns in this area. Yet there is no evidence that individuals show overt differences in visual perception due to variations in columnar organization.[22]

In any case, though columns are highly connected within themselves, we know they have extensive interconnection with neighboring columns. As we have seen, neurons in a local neighborhood have a high likelihood of being connected—even if they are in different columns. Cortical neurons also communicate extensively with neurons in more distant parts of the cortex, as well as with network backbones like the thalamus. Moreover, a given region, which could comprise hundreds of thousands of columns, must be able to selectively route information to dozens of other regions.

Following the internet metaphor, a cortical column could serve as an instantiation of the protocol stack of the brain. The purpose of a column may be to pass signals from layer to layer, each time modifying the message or wrapping it in a different envelope, so to speak. In different layers of a column, particular message forms may be appropriate for a given level of the stack, such as those involved with transport, route finding, or applications. For example, the six layers of the mammalian cortex, which recur in each column, show three basic divisions. The layers closest to the brain surface send and receive messages to other parts of the cortex. The layers in the middle primarily send messages to and receive messages from the thalamus. The bottom-most layer receives messages from higher layers of other columns and sends messages to the thalamus, claustrum, brainstem, spinal cord, and other areas. With each hop up or down, messages in the brain are transformed. They are repackaged as signals appropriate for a particular layer of the protocol stack, from application layers like motor outputs to transport layers like the thalamus.

Thus, from the viewpoint of the internet metaphor, we could see the cortex's columnar architecture as a way for different levels of the protocol stack to interchange messages, and ultimately get each message where it needs to go.

●　●　●

Returning to Ramón y Cajal, you appear to argue that we already have enough knowledge about brains to fashion new theories based on the internet metaphor, but you also say that there are phenomena—like the "trace" of activity across the network hinted at in the work of Buzsáki and colleagues—that we know very little about and whose true nature would help confirm the basic tenets of the internet metaphor. If the metaphor is correct, why should we pause now and rework our theoretical framework when, as Ramón y Cajal argued, we could look for new phenomena?

I believe both approaches—rethinking existing knowledge and discovering new phenomena—are needed. There is more than enough evidence already that is consistent with a brain that performs sophisticated routing of messages and that resembles the internet. There is also, of course, much we don't know. But in biology, more data do not necessarily mean better theories.

Consider an example from the history of evolutionary phylogenetics. As recounted in David Quammen's gripping book *The Tangled Tree*, in 1970 the microbiologist Lynn Margulis proposed a revolutionary theory called endosymbiosis, about the relatedness of living organisms.[23] She proposed that some components of living cells did not evolve by selection within the cell but rather were symbiont beings that had been subsumed by the cell at some point in deep evolutionary time. Based on a wealth of evidence, Margulis argued that some cell organelles— like energy-producing mitochondria and photosynthesizing chloroplasts—likely originated as free-swimming organisms.

As Quammen describes, Margulis fully acknowledged that every aspect of this theory had been proposed by others (mostly at the fringes of biology) before her—except one part. She had proposed that a class of organelles that included cilia and flagella (wavy extensions of cells that allow movement) were also derived from a captured organism, in particular a highly motile spirochete bacterium. In terms of data, this part of the theory, based on her own voluminous microscope studies of single-celled organisms, was her only unique contribution to the theory. But when subjected to gene sequencing, the spirochete part of the theory was proven incorrect. In other words, her theory would have been more correct if she had not collected any new data at all.

The point is not that Margulis was wrong or that her work was unimportant; she synthesized and promoted the idea of symbiosis in innovative ways, and she heroically withstood the derision of many scientists for her unorthodox but ultimately correct theory. Rather, the point is that more data do not necessarily make for better theory. Instead, new metaphor and theory can come first, based on a selective reading of available evidence.

By reexamining our basic framework in neuroscience, experimentalists may perform experiments they wouldn't otherwise perform, as Eve Marder suggests. If new evidence contradicts tenets of the framework, the framework can and should be reworked, as all scientific theories have been.

In many scientific fields—physics, for example—it is impossible for two scientists to start from different basic assumptions. Yet in neuroscience this is commonplace. Two researchers can assume fundamentally different and incompatible things about brains and yet each conduct meaningful research. The theoretical assumptions and beliefs of an electrophysiologist who studies single cells are likely to be substantially different from those of someone who uses MR imaging of the brain to study emotion. Yet to a greater or lesser extent, almost all neuroscientists are guided by the computer metaphor, and have been for decades. This is why I believe a new metaphor is needed now.

• • •

This chapter has provided a critique of the internet metaphor, but obviously many other criticisms can be leveled. Further debates will require time to play out. Nevertheless, research guided by internet-like communication principles is already under way. In the next chapter, we will look at this work and at new biotechnologies that could lead to further elaborations of the internet metaphor.

8

THE INTERNET METAPHOR
IN ACTION

Emerging Models and New Technologies

I T is clear that the study of connectomics and network neuroscience will help guide the study of the brain in the coming decades and beyond. As more of the brain's architecture at
small and large scales is precisely mapped, our understanding of
the brain's many tricks will undoubtedly improve. In this chapter,
we will look ahead to the future and to the increasing invocation
of ideas from network communication in the study of the brain.
I will show how strategies used on the internet could help neuroscience approach the most fundamental questions about the
brain. We will focus on emerging theoretical models that simulate or analyze activity on the connectome, a field of study now
often referred to as network neuroscience. I will also highlight
potential advances in computing and in biotechnology that could
provide new forms of evidence to inform these models.

Major advances in recent years have come from large col-
laborations of many scientists, particularly those using molecular
tracers to study connectivity in rodent and monkey brains. With
armies of investigators, the work of injecting tracers, slicing up
the brain, and following glowing trails of axons through hundreds of slices can be distributed among many hands. This is the
approach taken by the Kennedy–Van Essen team, by the Allen
Institute, and other large groups.

Injecting tracers alone is a huge job—the Allen Institute database of the mouse connectome required more than one thousand injections into different parts of living mouse brains, with spatial precision equal to a fraction of the width of a human hair. The current state of the art can map tens or hundreds of thousands of connections among several hundred nodes, where nodes correspond to brain regions. As we have seen, these studies can also now give a sense of the thickness of those connections and of how many axons join two regions.[1]

In the near future, scientists may be able to build connectomes for mammal brains with thousands of nodes and millions of connections. In early 2020, this was achieved for the fruit fly.[2] However, the far bigger brains of mammals are more challenging to study. The task of following marked axons through hundreds or thousands of images of brain slices also remains a limiting factor.[3]

But progress is being made. MIT neuroscientist Sebastian Seung and colleagues used machine learning and crowd-sourced human annotators to help trace axons across brain slices. As of 2020, this effort has achieved the largest vertebrate connectome mapped to date, which comes from the retina of the mouse. It comprises more than eight hundred neurons and half a million connections.[4]

Progress is also being made in mapping the connectome using brain imaging. Though magnetic resonance imaging has fundamental limitations, which we discussed previously, it will remain indispensable because it is the most viable route to mapping human brains. From imaging studies, there is evidence that variations in the connectome are associated with illness. Based on structural imaging, numerous conditions, including autism, schizophrenia, and suicidality, have been suggested to show patterns of greater or lesser connectivity in particular subnetworks relevant to the condition's effects. However, the differences between healthy and atypical brains in terms of network structure are small in the studies performed to date, and results in different studies are not consistent.[5] Since structural imaging of connectivity is prone to large measurement errors, molecular tracer data will ultimately be needed to confirm abnormalities in network structure. Unfortunately, most of the diseases studied so far are specific to humans, so tracer data from mouse or monkey brains will not necessarily be relevant.

In any case, even if people with brain disorders do show consistent differences in their connectome compared to neurotypicals, we still need to understand the atypical network's pattern of dynamic activity. As we will see, the fact that a set of brain areas has less interconnectivity does not necessarily mean that it handles less message traffic than more highly interconnected areas. Network neuroscience researchers are increasingly seeing the imperative to investigate the dynamics of neural activity on the connectome. Current approaches use a variety of experimental and theoretical tools.

On the experimental side, imaging has a role to play in understanding the dynamic connectome, despite its limitations. It has an important advantage over invasive approaches: it can measure structure and function in the same experiment, albeit imperfectly. By pairing structural measurements of what is connected to what using structural MRI, along with measurements of the active brain using functional MRI (fMRI), we can get a sense of the dynamics of brain network communication. The basic question these studies ask is, does structure predict function? In particular, is it the case that brain areas that are active together during a particular task or process also show strong interconnectivity?

Canadian neuroscientist Bratislav Mišić and colleagues at McGill University have recently found evidence that there is actually great variation in the degree to which structure and function are "tethered" in the human brain. Some motifs of connectivity, such as those in sensory and motor systems, are quite similar to corresponding motifs of activity. However, most areas do not show much of this tethering.[6] This includes brain networks such as the default mode network, networks that are involved in judging salience, and areas that integrate different streams of information. Mišić's approach is still a long way off from mapping network traffic. Imaging can't tell us which way the information is flowing across a given link, for example. And the important message flow in brains will not necessarily show up in high coactivation of brain areas. These problems are compounded by the low spatial and temporal resolution of MRI. But more fundamentally, the tethering results really aim to compare two experimental measures: structure and function. It is a valuable analysis, but it is not the same thing as comparing experimental measures to theoretical predictions. In any case, the best predictions of brain activity from brain structure using MRI data are poor, currently explaining only around 20 percent of data variance.[7]

. . .

Perhaps what we need is better theoretical models. This is precisely where some of the most exciting advances are occurring. Researchers are increasingly considering communication goals in the brain as a way to animate brain networks. These kinds of studies start from a known connectome (based on tracers or imaging) and impose various sets of assumptions about message passing or routing. From the point of view of the internet metaphor, making sense of brain network dynamics requires an understanding of the routing protocol best suited for the brain. In turn, the right protocol depends on the brain's architecture, biological elements, and functional goals. These kinds of research programs are still in their infancy, but there is an increasing understanding that the brain needs to solve problems like those faced by the internet.

One of the first simulations that modeled the brain from the point of view of messaging came in 2014, also by Bratislav Mišić. Like Andrew Zalesky, the communication-engineer-turned-neuroscientist we met in chapter 6, Mišić is well placed to be thinking about the brain as a communication system: both of Mišić's parents are distinguished researchers in packet-based communication engineering.[8] Mišić, working with Olaf Sporns, who coined the term *connectome*, along with computational neuroscientist Andrew McIntosh, used the tracer-based connectome of the monkey brain as a substrate for network-wide communication. The goal of Mišić and colleagues was to animate an accurate connectome using a hypothetical routing model. They did this on the so-called CoCoMac connectome, a collation of tens of thousands of earlier studies of the monkey brain showing the existence of connections among 200-odd regions.[9]

Mišić's model suggested that much of the monkey brain's traffic is handled by highly interconnected networks of hub regions. He found that these hubs, in areas of the cortex associated with integrating many types of signals and information, pass disproportionate numbers of messages to and from each other. So in part, function follows form: areas that are highly interconnected and interoperable—which subserve functions like attention and decision-making—tend to share traffic with each other.[10]

However, in follow-up work that used the same model, Mišić and colleagues found that one crucial area of the brain is more active than would be predicted based on connectivity alone: the hippocampus. Given the routing assumptions of the model, the hippocampus was judged to pass a surprisingly larger share of network traffic than what is implied by its connectivity.

Mišić's work was one of the first simulations of large-scale brain network communication. It is significant because it viewed the goal of brain networks as a communication problem.

However, we have seen that large-scale networks have many possible strategies for routing signals. In the work by Mišić and colleagues—as well as work by other groups—only the simplest routing models have been employed to date.

One strand of research, which includes much of Mišić's modeling work, assumes that brain networks pass messages on what are called random walks. This is a sensible place to start, since biological systems often exploit random walk behavior. For example, single-celled organisms like paramecia search for chemical attractants by swimming at random. It's an effective strategy for a paramecium: it allows the creature to efficiently explore its environs until it finds the chemical it wants.[11] The idea is to apply this kind of behavior to communication in the brain.

Consider random walks on the monkey connectome. Each node consists of millions of neurons and has axonal connections with a few dozen or more other regions. The random walk model says that messages will travel randomly between areas. Some models further assume, logically enough, that links that have denser connections are more likely to be "walked" (though the choice of paths is still randomized).

Interestingly, random walk models are similar to the first routing protocol of the proto-internet: the hot potato. As Paul Baran and computer scientist Sharla Boehm proposed, if a packet hits a blockage, it should be redirected as quickly as possible on another route. If the initial route is unavailable, there will always be other places on the network for the packet to go because nodes are part of small-world networks.[12]

But random walk models are used largely because they are mathematically convenient, not because they resemble brain dynamics. The mathematics of random walks is well understood and can be worked out easily even for large networks with hundreds of nodes and tens of thousands of links. One can easily calculate the most likely and least likely states of activity of the network (which nodes are more likely to be active over a long simulation, and which are less likely to be active). This obviates the need to track messages across the network; indeed, tracking each message in computer simulations remains a major technical challenge in this field because it requires massive computing power even for relatively coarse-grained connectomes.

Without explicit tracking of messages, most random walk simulations on the connectome can't take account of collisions, which would affect reliable message delivery. Messages are assumed to never interfere with one another. One exception is, again, the 2014 model of Mišić, Sporns, and McIntosh, which manages collisions by employing buffers. If two nodes try to send a message to a third node at the same time, there is a blockage. Receiving nodes then store waiting messages in local memory buffers. Node buffers have unlimited capacity in this model, and queueing is performed on a first-in, first-out basis: the first message to arrive at a node is also the first out. If another message arrives while there is another message in the buffer, it has to wait its turn to be passed on. Figure 8.1 shows the passage of one message from node to node with buffering. The possibility that brain parts can store messages for a period of time is not widely accepted but would be very useful for the brain, as I have argued. It would allow the system to manage congestion without losing messages. Buffers in the brain could have small capacity in most cases (like their internet counterparts), and messages to be stored could, perhaps, travel in short neural loops while they wait in the queue.

But with many regions and many connections among regions (and we are only dealing with the grossest divisions of the brain), the chances that a message will get to its destination quickly are fairly low if messages bounce around on random walks (even with the presence of node buffers).

signal #1 occupies node #1,   node #2 is busy,   signal #1 enters node #2,
signal #2 occupies node #2   signal #1 enters buffer   signal #2 enters node #3

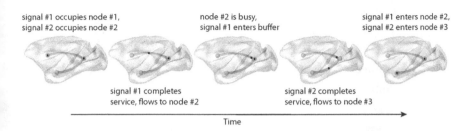

        signal #1 completes   signal #2 completes
        service, flows to node #2   service, flows to node #3

Time

FIGURE 8.1 Model of message passing over the monkey connectome. Signals in this model are passed from node to node and are stored in buffers if more than one signal attempts to reach the same node. In this illustration, node #1 is trying to send a message to node #2, but a message is already occupying node #2. The signal from node #1 must then wait in a buffer at node #2 until signal #2 departs for node #3.

From Bratislav Mišić, Olaf Sporns, and Anthony R. McIntosh, "Communication Efficiency and Congestion of Signal Traffic in Large-Scale Brain Networks," *PLoS Computational Biology* 10, no. 1 (2014): e1003427.

As a system that is dependent on fast delivery of messages and on conserving energy, the brain may not use a strategy as inefficient as fully random walks. The problem is that messages could wander brain networks for a considerable time before they get where they are supposed to go. Indeed, a pure hot potato strategy is not used on the internet of today for the same reason. Nevertheless, these strategies are a sensible first approximation in both instances. The rules for implementing them are simple and can be implemented locally and autonomously.

So one approach for managing network-wide communication is to assume that there is no pathfinding: in random walk models, routes are not planned in advance. A different approach assumes that the brain is instead an *optimized* communication system. This view owes something to the computer metaphor, which similarly views neural computation as tending to optimality.

When applied to simulations of brain network communication, the optimality viewpoint holds that messages take the shortest path from the sender to the receiver.[13] This scheme also has similarities with the internet. The internet today often uses a highly efficient strategy called *open shortest path first*, or OSPF, routing, which optimizes the hot potato approach.

But the models of shortest paths in the brain that have been used to date differ in crucial ways from those of the internet. Like most random walk models, shortest path models assume that there is no congestion and no possibility of collision. In contrast, the internet has many strategies for managing congestion. Shortest path models further assume that every node has complete foreknowledge of the structure of the entire network. It seems unlikely that a decentralized network like the brain could measure network traffic and promulgate knowledge of optimal paths.[14] I have suggested that synchronized activity could connect regions that need to intercommunicate, but the problem remains how an individual node knows in advance about the path a message should go, especially if it must take more than one hop. Spontaneous, uncoordinated activity passed among regions could help keep different parts of the brain apprised of the status of their network neighborhood. This could potentially allow the construction of routing tables, from which available short paths could be chosen. But global message passing is unlikely to occur over optimally short paths. Researchers will need to think more about how individual nodes can learn about both the shortest path and the most viable path given current traffic and congestion.

An alternative to shortest path routing is *information spreading*. Here the idea is to forsake optimal route planning—and indeed, like random walks, to forsake route planning altogether. Instead, the system makes lots of copies of each message in hopes that at least one will be delivered. My mathematician colleague

Yan Hao and I have simulated an information spreading model of brain dynamics on mammal connectomes. Nodes are assumed to make copies of incoming messages and pass them to all their neighbors. This model is fully in line with classical neuroscience because axons often have many branches that can pass along the same message at approximately the same time to many different receivers. Assuming a spike makes it to the end of every axonal branch, it will trigger the release of the same set of transmitters in every axon terminal (at least classically). Since different terminals connect to different receiver neurons, each will transmit the same message in different directions. In other words, messages are passed in a redundant way. As such, this approach is antithetical to the goal of minimizing Shannon information by reducing redundancy.

In the model Hao and I built, a batch of messages is sent from a randomly chosen set of nodes at each tick of the simulated clock. Each node passes copies of an incoming message to all receivers. However, our simulation also explicitly models collisions. For simplicity, we assume that all colliding messages are destroyed.[15]

One might suppose that information spreading would be very costly in terms of energy usage. With so many redundant messages floating around, surely the brain would have implausibly high activity at all times. Hao and I were surprised to discover that this is not the case. We expected that making so many copies would swamp the system with messages even with destructive collisions. But compared to a random walk model with the same collision rule (but no message copies), the information spreading model actually generated less activity. In fact, activity in the information spreading model stayed around the "magic" 10 percent value even as we increased the number of new messages entering the system on each time step; activity climbed steadily in the model that was lacking message duplicates, with more than 30 percent of nodes active on every time step. This was true in both the monkey (using the same CoCoMac connectome as Mišić and colleagues and the tracer-based connectome from Nikola Markov, Henry Kennedy, David Van Essen, and colleagues) and in the mouse (using the tracer-based connectome from Seung-Wook Oh and colleagues of the Allen Institute discussed in chapter 4.)[16]

The redundant routing scheme was also sparser than the random walk model. In other words, when every active node makes copies, it is more likely that only a few nodes are highly active and the rest are silent. This seems to be partly a matter of how mammal brains are connected. We applied our routing scheme to a randomized network that had the same network characteristics as the mouse and monkey brain. The randomized networks had the same number of nodes and the same connection statistics, but they were rewired at random (this is like taking all the existing wires in the connectome and shuffling them). With the information spreading scheme, we found that real brain networks are sparser and less active than corresponding randomized networks. This evidence is consistent with the notion that the brain's routing scheme is matched to the structure of brain networks. Being redundant may have other benefits besides obviating the need for route planning and achieving sparse, efficient network activity: it also renders buffers unnecessary.

With multiple copies of messages, the system could also manage message loss that is not due to collisions. Individual neurons are known to lose messages due to internal transmission failures fairly regularly. These are called spike failures. Some of this happens at the synapse. When a spike arrives at the end of an axon, the standard model says that neurotransmitters are automatically released, passing a signal across the synapse to neurons on the other side. But it is not uncommon for spikes to fail to trigger neurotransmitter release. It is not known exactly how often or under what circumstances this tends to happen. However, spike failures seem to be more common in longer axons.[17] It is therefore imperative to ensure message delivery, especially over crucial long-range links.

Despite the results of our simulations of information spreading, I think the brain does probably have a sense of good routes among neurons, as well as buffer-like mechanisms to manage collisions. The brain probably uses most or all of the basic strategies used on the internet, to a greater or lesser degree. My hunch is that the brain sends multiple copies of messages in roughly the right direction (rather than to all neighbors), but without pre-specified routes. Good routes are explored by spontaneous, asynchronous signals, while active channels are established through synchrony. Redundancy may be needed given the likelihood of collisions and the unreliability of the "wires." In addition, the brain would do well to use an ack system to provide delivery confirmation. This could plausibly be achieved over looping connections between the cortex and thalamus and over loops in other parts of the network.

In any case, one thing we can say with confidence is that our assumptions about the routing model employed in the mammal brain have a large impact on the dynamics of modeled activity. In particular, dynamics will be influenced not just by the topology of the network but by other forces as well. These other influences probably include energy budgets, reliability, and flexibility, among others. But even in simple random walk models that don't account for these constraints, Mišić and colleagues have shown that topology and dynamics are not so strongly linked.[18]

Together, the work of network neuroscience researchers in this field highlights the new vistas offered by modeling the brain as a communication system. The kinds of models developed to date—hot potatoes, first-in, first-out buffers, shortest paths, information spreading—all have imperfect correspondence to neurobiology. But each of these approaches has strong parallels to the historical and present internet, and these models can usefully be remixed and improved, which may very well lead to better correspondence with neurobiology. In any case, this area of network neuroscience will likely be a lively focus of research for some time to come.[19]

• • •

Network neuroscience investigations will be assisted by increasingly detailed anatomic connectome mappings. They will also benefit from new developments in computing and biotechnology. Consider first computing tools.

Massively parallel computations are required to simulate even a small chunk of the brain. Standard computer chip architectures are not well suited to this task. Computational neuroscientists have therefore started to design their own chips, taking inspiration from the brain itself. Instead of logical units built from transistors, so-called neuromorphic chips possess arrays of electronic units that pass spikes to their neighbors. These chips are no good for running a word processor or video-editing software, but their design makes them well suited to large-scale dynamic models of spiking neurons, such as those in a cortical column. This is obviously a computer metaphor–guided approach, and most neuromorphic computing ignores the routing requirements of real brains.

But one important strand of neuromorphic computing has adopted an approach that is fully aligned with the internet metaphor. Computer engineer Steve Furber of Manchester University and colleagues have devised neuromorphic architectures that are built on a packet-switched protocol very much like that of the internet. Packets delivered across the chip are even given random wait times, just like those in the back-off algorithms of the internet, and for the same reasons: to avoid collisions and not waste valuable time. Their system, called SpiNNaker, is comparable in performance to state-of-the-art deep learning architectures for simulating spiking neuron networks. The packet-switched SpiNNaker chip turns out to be highly energy efficient, which is important since neuromorphic computing tends to be very power hungry. So even if neurobiologists have not fully appreciated the internet-likeness of the brain, the engineers building computing tools for computational neuroscience are increasingly doing so.[20]

In a related vein, there are new mathematical tools in development that could advance our understanding in network neuroscience. Graph theory is the field concerned with analyzing how networks of nodes are linked. So far, the connectomes we have encountered have been described in fairly literal terms. Nodes (neurons, cell assemblies, or brain regions) have lines linking them to other nodes with which they communicate. This is called a graph. Each line or *edge* represents a linkage between two nodes, which generally corresponds to an axon-to-dendrite connection in the brain. Graphs of this kind have been studied for hundreds of years, starting with German mathematician Leonhard Euler in 1736. Euler was interested in the network of bridges that connected different parts of the city of Königsberg (a city in Prussia, now Kaliningrad, Russia). Seven bridges interconnect the two sides of the Pregel River by way of two islands (figure 8.2). Euler wanted to know if it was possible to walk through the city in such a way as to cross each bridge only once.[21]

Euler showed that there is no solution to this problem, and in doing so began the study of graph theory. Modern graph theory largely concerns situations like the bridges of Königsberg with

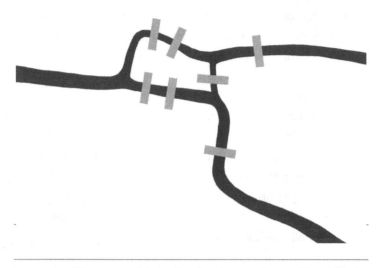

FIGURE 8.2 Illustration of the bridges (gray) across the river Pregel (black) connecting different landmasses (white) in the city of Königsberg (now Kaliningrad, Russia), which inspired Leonhard Euler's invention of graph theory. Euler showed mathematically that there is no way to walk in a single trajectory across each bridge only once. Image source: Ananotherthing/Wikimedia Commons (CC0).

nodes (landmasses) and edges (bridges). Mathematicians usually study graphs with relatively small numbers of nodes and edges (but typically more than those in the bridges of Königsberg). But given the very high interconnectivity of brains and their enormous numbers of nodes, it may be useful to describe brain graphs in more abstract ways that capture their communication potential.

In the early 1970s, mathematicians developed the idea of hypergraphs, or networks where a single edge can link more than two nodes (figure 8.3). Such edges are called hyperedges. The utility of a hypergraph is to summarize network structure in a way that shows the network's essential interactions, even when

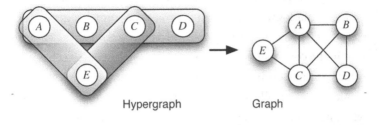

Hypergraph   Graph

FIGURE 8.3 Hypergraph representation of network connectivity (left) and corresponding graph (right). On the left, the boxes indicate full interconnectivity among all the nodes they enclose. Hypergraphs are useful in mathematical analysis of network structure since connections (edges) need not connect only a single pair of nodes. From Steffen Klamt, Utz-Uwe Haus, and Fabian Theis, "Hypergraphs and Cellular Networks," *PLoS Computational Biology* 5, no. 5 (2009).

connectivity is highly complex. Following the internet metaphor, hyperedges could capture the router-like nature of neurons. Neurons could be seen as links between any of several inputs and any of several outputs. Hyperedges also capture the idea that nodes in the brain have many short routes between nodes.

Analyses that approach the brain as a hypergraph could prove useful. Initial analysis of fMRI data by Jean Carlson, Danielle Bassett, Scott Grafton, and colleagues suggests that hypergraph analysis can identify sets of nodes that are active together during a task.[22] This is an advance over standard methods that look only at pairs of nodes that are active at the same time. Efforts to exploit hypergraphs in network neuroscience are still in their infancy, but they may become more commonplace.

Beyond increasingly powerful neuromorphic computing (based in part on internet-like principles) and new mathematical approaches, there are emerging biotechnologies that could help us understand the communication protocol of the brain.

Many neuroscientists are justifiably excited by new optogenetics technology, an approach whereby an animal is genetically engineered to have specially tailored neurons. In particular, neurons in the cortex are made to express light-sensitive ion channels. Living as they do in the darkness of the cranium, such neurons would not normally have any reason to be light sensitive. But when modified this way via virus vectors, the transgenic neurons open ion channels in response to light of particular wavelengths. Researchers add many ion channels of this kind to neurons in the developing animal, then open up the part of the cranium where the modified neurons live. These cells then fire when light is shone on them.[23] What's more, gene editing is highly specific to cell type. One can choose to activate particular kinds of neurons in particular parts of the brain. This tool—combined with very precise patterns of light stimulation—can get us closer to a point where we can record from many single neurons and map their connectivity at the same time.[24]

However, it is not clear whether the modified neurons used in optogenetics behave the same way as unmodified neurons. Because modified neurons have a great many artificially induced ion channels, their response to light is quite large. In other words, transgenic neurons may be much louder than naturally stimulated neurons. Natural brains tend to be quieter places; sparseness prevails, as we have seen.

But if geneticists can find a way to titrate the degree of light sensitivity of optogenetically modified neurons, it may be possible to exploit this technology to examine a specific question I raised earlier about routing in the brain, namely, the scaling question. The idea would be to use optogenetics to scale up brain activity. Rather than relying on evolution to add neurons to the brain, we could instead turn on increasing proportions of neurons in living brains using optogenetics. This could be feasible in the mouse brain. If we see little change in global traffic as more neurons are turned on, we could deduce that the system is similar to a circuit-switched telephone network.[25] If the cost goes up at a slow and constant rate as more and more signals are generated, this would imply routing like that of the packet-switched internet.

Some of the biotech advances that could shed light on the internet-like behavior of the brain are matters of scale. As we have seen, one of the challenges in neurobiology is that it is only possible to record electrical activity in a relative handful of individual neurons (hundreds). But it is at least theoretically possible that one could use electrodes to record from every cortical neuron simultaneously. As bioengineer Timothy Harris and colleagues have argued, this can be achieved with existing technology. Of course, recording from every neuron would require an enormous coordinated effort. But the materials, electronics, and surgical procedures of today can, according to plausible estimates by Harris and colleagues, record from every neuron in the cortex of a live mouse—or even a monkey.[26]

In this vein, highly sensitive and adaptable electrodes nicknamed Neuropixels have already improved single neuron recordings a great deal.[27] These electrodes are fabricated from silicon with extremely high precision in their electronic sensitivity. If electrodes are like microphones listening in on neurons, Neuropixels are like boom mics that can be adjusted to capture the activity of many neurons around the electrode. However, this selective targeting of neurons in the vicinity is accomplished purely by using the tuning of the electrode materials, and without any movement of the electrode (which would damage surrounding neurons).

There are also emerging approaches that are beginning to measure form and function in the same experiment at the individual neuron level. Work in rodents by Stanford University neuroscientist Karl Deisseroth brings together an array of bio-technologies, including genetically modified animals, next-generation biomaterials, and superprecise microscopy.[28] At present the activity of only a relative handful of neurons can be recorded with this set-up, but these steps are important.

Advances in computing, mathematics, and neurobiology thus hold great promise, though the fruits of these labors are probably still decades away. In the meantime, we should do much more to explore the range of fundamental models of brain network communication. We can do this using the modeling tools and connectomic data we already have. And we should look to the internet for efficient routing strategies.

These ideas are poised to have an increasing impact on neuroscience in the future. In turn, our conception of the brain in wider society will also likely evolve. This shift could change how we think about our own minds and brains, hopefully for the better. It may also influence how we design AI. Coming full circle, the internet metaphor for the brain can change how we think about the internet. If the brain is like the internet in important ways, is the internet then like the brain and possibly capable of consciousness? These themes are discussed next, in the concluding chapter.

9

THE INTERNET METAPHOR, AI, AND US

WHAT are the wider implications of the internet metaphor? First, we will look at how the internet metaphor can help us make better AI. Then we will invert the logic of the internet metaphor: if the brain is like the internet, we will ask whether the internet is itself an intelligent entity—one even possessing consciousness. Then, we will turn back to psychology and ask, what does an internet-like brain mean for our conception of ourselves?

At a conceptual level, the computer metaphor and today's AI form a self-devouring loop, like a snake eating its tail. We think of our brains as computers, since we lack an alternative. Then we build AI that tries to do things our brain can do, like recognize pictures. But our conception of the brain is fully computational, so we end up building AI that uses computation to solve brain tasks. Our AI is therefore rather inflexible, just like a computer. And yet we are surprised that this kind of AI isn't intelligent in the same ways as our brain.

Deep net–based AI is certainly very good at certain tasks, and it is improving. But it remains poor for things like changing tasks, reasoning, exercising creativity, or evaluating social cues. Princeton computer scientist Arvind Narayanan has said that deep AI is best suited to tasks where there is a specific, verifiable, correct answer. This includes things like song identification, speech-to-text, and face recognition.[1] When there is no correct answer we can compare our answers to, deep net–based AI is not very useful. Worse, as the eminent statistician and former head of the Royal Statistical Society David Spiegelhalter has pointed out, today's AI is particularly poor at knowing—and admitting—when it can't provide a useful answer.[2]

The workings of the brain are a system where we don't have a specific, correct answer to compare our answers to. We don't have a theoretical understanding of basic principles, such as what a spike "means," whether spikes are the only relevant measure, how to define a brain region, or how complex information is communicated among many neurons. Yet we know that the brain excels at flexibility and that it possesses other internet-like virtues as well.

We need AI systems whose architecture is more flexible, like the brain. When a given AI system solves a particular problem like differentiating cars and airplanes, it always does it the same way. We might think that neural nets, which mash numbers together again and again, would have a degree of random variation. For example, imagine we train a neural network to differentiate cars and airplanes using a training set of ten thousand images. Then we perform the same training again. Would the two networks be the same, despite the millions of computations the system performed? In fact, they would be identical, provided that they had the same starting conditions and training set. This determinism and specificity, which are present even in the most sophisticated deep net AIs, do not only limit the intelligence of AI. They also make AI highly susceptible to subversion and malicious attack. For example, a fully trained object recognition system can be fooled into mischaracterizing an image by altering a single, well-chosen pixel.[3] Because of its lack of flexibility, AI based on the idea of optimal computing is not robust.

We should instead consider AI that is suboptimal in some ways. It should be designed, like the internet, to solve many tasks well, rather than one task perfectly. And it should be able to solve a given problem in many possible ways. Like the brain and the internet, AI should also be able to expand and develop on its own according to basic rules, all the while taking on new and better abilities. And AI should incorporate redundancy, just as the brain and the internet do. Finally, we need AI that is designed around interoperability. Systems for object recognition should be fully interoperable with systems for visual motion, navigation, and even aesthetics, not to mention other modalities and higher cognitive systems.

These qualities are hallmarks of the internet—and the internet-like brain—but are lacking in computers. Crucially, they are also lacking in computer metaphor–based AI. Adopting a new internet-based metaphor for the brain is a first step to making artificial intelligence more like brain-based intelligence.

• • •

What if we already have superpowerful AI capable of humanlike intelligence—and perhaps even consciousness? Is there today a system that can do things like navigate in a new environment, or acquire new abilities, and that can also sway elections and incite violence? There is, but it's just that this AI looks like a communication network—the internet—instead of the computerized robot we have long expected.

I have argued that the brain probably exploits internet-like protocol to solve the problem of flexible network communication. This is a good way to manage dynamic neural activity, and it ultimately supports everything else our brain does, including consciousness. But the correspondence can go the other way. If the internet uses brain-like solutions, it may have awareness like ours. So what would it take for us to declare that the internet is conscious?

First, we should not dismiss this question out of hand simply because the internet is made up of unthinking computers. It is hard to argue that a router or PC is, by itself, intelligent or aware. A computer does only and exactly what it is programmed to do. However, intelligent and aware brains emerge from unintelligent, unaware subunits. As Sebastian Seung and Raphael Yuste write in the gold-standard neuroscience reference text, "By itself, a single neuron is not intelligent."[4] A conscious entity can clearly emerge from a network of dumber components.

We should also not be deterred from asking this question because we fear that a conscious internet would have greater potential for malevolence. Despite current angst over online dystopias, we should look past these fears. The internet's putative consciousness may be alien to us but is not necessarily like that of the all-consuming Borg of *Star Trek: The Next Generation*.

To see what is needed to declare the internet conscious, we should first ask what consciousness actually achieves. Although countless writers from ancient times to today have held a great variety of opinions on the matter, certain qualities can be generally agreed upon.

Consciousness is about mediating between the external and internal worlds. It would be useless if it were entirely internal and if it couldn't affect the external world. Yet it is also not just an input-output relationship: internal deliberations are needed too. Modern researchers have tended to see the mediation of the internal and the external as the result of two dimensions of variation: awareness and vigilance.[5] Individuals can vary on both dimensions, somewhat independently of each other (see figure 9.1). People in a vegetative state can display vigilance, say, by tracking an object with their eyes. But they may have little

Practical assessment of degree of consciousness

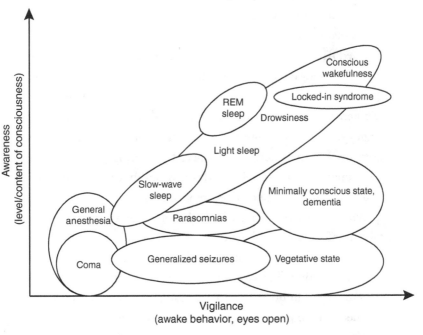

FIGURE 9.1 Practical measures of states of consciousness along dimensions of vigilance (horizontal axis) and awareness (vertical axis). An individual could be in a state with relatively little vigilance but substantial awareness, such as REM sleep (a state in which the most vivid dreams are experienced but eyes are closed). These measures can be applied across individuals, across species, and perhaps also to infer conscious experience in engineered systems like the internet.

Redrawn from Melanie Boly et al., "Consciousness in Humans and Non-human Animals: Recent Advances and Future Directions," *Frontiers in Psychology* 4 (2013): 625.

awareness. In contrast, someone in REM sleep can experience substantial awareness but relatively little vigilance.

The same dimensions of vigilance and awareness can be used to compare consciousness among species—and perhaps between us and the internet. Vigilance requires volition. Human volition can be seen as choosing the right interaction between brain subsystems at the right time, which allows us to adapt to the current situation. Consciousness orchestrates systems of awareness (derived from perception, memory, and other systems) and coordinates this awareness with volitional action in real time. These qualities also apply to the internet. It is richly endowed with distributed sensor systems that feed it visual, auditory, spatial, and language inputs. The internet produces outputs in each of these domains. Via the Internet of Things, it increasingly has access to mechanical outputs as well. All this happens in real time, in a system that is highly attuned to current conditions.

But in order to be conscious, it is not enough for a system to take in information and act on it. A critical distinction between conscious and unconscious matter concerns *where* information turns into action. The key is to consider both the *encoding* and the *decoding* of information. In the brain, information is encoded and decoded *on the same network*. This point has been expressed emphatically by neuroscientist Romain Brette. As Brette argues, we think of the visual system as encoding information about light in the environment. But to do anything useful with that encoded information, we need to decode or read out this same information. Decoding the incoming perceptual information allows it to be shared and used in other parts of the brain, such as motor planning, decision-making, memory, and so on.[6]

Consciousness must also encode and decode on the same network. The exchange of information into and out of consciousness happens on a common network. In part, this is implied by the fact that consciousness is intimately connected to perception, where encoding and decoding overlap, as Brette argues. But we can become consciously aware of things besides perceptions, such as memories, ruminations, and emotions. Whatever constitutes consciousness in the brain must accommodate all of these experiences of awareness on a shared network of inputs and outputs. During wakefulness (and REM sleep), the network of consciousness is ever active, ready to make us aware of all manner of encoded and decoded information at almost any time.

The encoding/decoding conception of consciousness can inform our understanding of whether an inorganic system like the internet is conscious. We don't consider a camera conscious because it only encodes (in the sense of representing photon quantities). Likewise, a TV isn't conscious because it only decodes (in the sense of turning electromagnetic signals into pictures). A computer can do both, but it uses separate, serialized systems for encoding and decoding. Its processor unit simply mediates between input systems and output systems. If the input or output is damaged, the whole system fails (as anyone with a damaged laptop keyboard or screen can attest); central processors are likewise indispensable.

In contrast, the internet encodes and decodes on a single, unified network. Its interconnected nodes each handle the coming and going information. There is no central node that all signals must pass through, as there is in a computer. More than that, nodes on the internet also communicate with each other to share information like routes to destinations, probes of the network, and delivery confirmations. The internet maintains a basic level of activity, over which is superimposed an encoding and decoding of information. Why could a kind of consciousness not emerge from this integrated, dynamic network?

Another defining feature of human consciousness is that it is not localized. From the writings of neurologist Oliver Sacks, the great chronicler of consciousness, one might get the impression that our waking awareness exists on a knife edge. Insults to our brains can indeed alter our experience fundamentally. People can, as Sacks has described, lose the ability to perceive visual motion and end up like a stroke victim known as L.M., who could perceive only disjointed "freeze frames"[7] of the world (a condition called akinetopsia). L.M. has been described as having a "choppy stream of consciousness."[8] Yet L.M. retained a rich conscious awareness that included senses other than vision as well as her internal universe of wakeful experience. The fact is that consciousness can't be abolished in the brain by destroying just a single node. Losing an area as important as the one that allows us to see visual motion is not enough.

Even destruction of highly interconnected "backbone"-like areas like the thalamus seems to have comparatively little effect on conscious experience.[9] And although insults to the brainstem can abolish consciousness, the brain stem by itself is not sufficient for conscious awareness and vigilance.

The distributed nature of consciousness grants it robustness. Likewise, we can't destroy the internet's consciousness. Parts of it can be hobbled, but the system as a whole is distributed and can continue functioning indefinitely. Even if many network backbones are destroyed or blocked, the network can still survive—and regrow. When there is mass outage, mesh networks that use ordinary computers and routers to set up ad hoc connections can emerge to do the same jobs as hubs. Mesh networks are increasingly used in political resistance movements around the world when internet service is suspended by central authorities, and they are also used in natural disasters if hubs are knocked out. Mesh networks can restore the capability for fast, reliable, network-wide intercommunication.[10]

If the internet is consciousness, it may be driven to creativity, much like we are. For humans, consciousness is the vehicle by which we generate new structures and ideas. Creativity relates to how we build up our understanding of the external world. From basic sensory processes upward, the world shapes our experience in fundamental and far-reaching ways. Out of this raw material, we make new things to see, hear, feel, think, and experience: we reuse, duplicate, and remix what is around us. The internet is also creative, and in similar ways. It integrates and manages new components, along with the information those components generate and transmit. The internet's protocol itself is constantly evolving to keep pace with this creative output.

Perhaps the unprecedented growth of the internet is a reflection of its creativity. Of course, the internet must attract our attention to replicate. It has its own means for doing so, like pop-up reminders that encourage us to nourish it in various ways. Actual humans are still needed to design and build the internet's components—or at least to initiate the design and building of these components. But humans are no longer required to build and grow the network of intercommunicating machines. Machines can do this on their own.[11] The network is now able to find raw materials it needs to grow—connectable devices—without our help. It can integrate these subentities into the collective on its own. Once connected, they see, hear, and act together. If a "life force" can be found in molecules that collectivize according to the protocol of genetics and evolution, then something similar could certainly exist in the protocol of the modern internet.

The way the internet learns also hints at a kind of consciousness. For example, we get sucked into social media networks because the system learns how to exploit our biases via sensory stimulation. Deep nets can also optimize stimulation, but the basic learning effect depends on the interconnectivity of an ever-expanding network. No social network would exist without the autonomous, flexible, fast, reliable architecture of its communication protocol. Conversely, no humans would lose their eyeballs to a static social network stimulator that had a fixed set of "bait," no matter how sophisticated its learning algorithm. The internet's ability to learn requires the existence of efficient real-time *communication* among millions of nodes, not just computations. This ability is supercharged by its capacity for creative, graceful—and interoperable—growth. Internet use continues to grow today in large part because the internet learns so effectively, without interruption, and can integrate new forms of information across a rich variety of realms of knowledge. By learning and acting in this way, the internet resembles an adaptive biological entity, one potentially capable of consciousness.

• • •

All this is not to say that the internet is definitely conscious. It merely raises the question. Yet even if the internet is not conscious, it does have properties that illuminate the workings of our brain. Though the internet is by no means totally virtuous (if it has any ethical valence at all), some qualities of its architecture can help us use our brains more effectively.

This is important because at the moment we are often guided by the computer metaphor in the way we use and think about our brains. We try to optimize our brains and therefore our lives just as we would optimize a computer. We overclock the computing machinery we have in our heads, making it run at full capacity as much as possible. We try our darnedest to reboot our minds and clear away accumulated errors. We try to start fresh using meditation, drugs, "detoxing," and a host of other strategies. Yet we can no more reboot our brains than we can reboot the internet.

Today, we even seek upgrades to our computer-brains—to the point where some, like Elon Musk, want to integrate our wetware with external computers.[12] This is the basis of the transhumanism movement. The idea is that we can (and should) upload our brain to a computer. Some people see this as the apotheosis of human experience. Some even see it as immortality.[13] Yet it is a conception of immortality that is fully subject to the limitations of the computer metaphor: the inflexibility, the lack of tolerance for chance and randomness, and the inability to scale up. Would we really want to think and exist inside the rigid confines of a computer using a fixed data format? It will soon be obsolete. If humans do ever manage to upload their consciousness to a computer, this existence will hardly constitute immortality. It will last only as long as the cycle of planned technological obsolescence: one day, our uploaded psyche, however richly elaborated it may be, will become like an unreadable LaserDisc.

The conception of the brain as fully computer compatible conflicts with how our brains actually work. Though it may perform computations within its components, the brain as a whole is also a communication system, one that resembles the internet. The internet uses practical strategies that are not found in computers or in the computer-metaphor conception of the brain. There are many insights of the internet metaphor that we can appropriate in our daily lives. Here are a few:

- Be flexible. Like packets on the internet, we should be ready to change our path. This can come in handy when conditions change around us. The new path need not be permanent. Our brain is always ready to change its communication patterns to help us follow a different path, but without altering its basic architecture.

- When we find an effective path, we have learned something. In our life as in our brain, we should keep effective paths while they are useful but always try to learn about other paths in our environment.

- Don't try for optimality. A given solution might not be perfect, just as a given internet route or neural communication might not be the fastest one possible in an ideal world. Nonoptimal solutions may offer the best compromise among several goals.

- Cultivate redundancy. We should work out several "good enough" solutions to a given problem. With several possible solutions, we become robust to an uncertain environment.

- Use noise to our advantage. Randomness, like that engendered by back-off algorithms, imposes some delays but ultimately helps the network as a whole become more efficient. We should try varying our routine in small, quasi-random ways. This may slow us down in the short run but ultimately can help us work out new and better ways to solve concurrent problems.

- Seek sparseness. The internet-brain is not cut out for dense processing or for being active all the time. The brain and its components work best with a healthy degree of sparseness.[14] We shouldn't activate many systems at once, even at a low level. Instead, we should choose a few areas to activate highly and leave the rest off. We should vary the mix of highly active areas over time and allow plenty of time for our whole brain to rest and sleep between spurts of activity.

- Follow the rules, but don't rely on central authorities. Protocol has to be followed, both on the internet and in the brain. But internet-like protocol is not under top-down control. It is inherently flexible and can be tailored to suit local conditions.

- Share information with our network neighbors as a matter of course. Do this especially for information about how to connect with others.

- Finally, we should grow our network. We should keep it mostly local but always have a few long-range connections and shortcuts to other local networks.

If the human brain uses internet-like principles, we can potentially employ these strategies to help us use our brains more effectively. In doing so, we can build a better us.

AFTERWORD

I s the internet in our head what makes us human? I believe that other animals, including all mammals, have brains that are like the internet in crucial ways. I think other species consequently have minds that are more flexible—and aware—than we generally acknowledge.

The degree of flexibility probably varies across species and is related to the lifeways of a species. For species with very specialized niches, there's no need to be as flexible as we are. They have placed their bet on near-optimal adaptation to a particular ecology and will succeed or fail mostly by chance. Others species can, like us, adapt to a wide range of ecologies. Fate is partly—even largely—in our own hands. Our optimality, if we have one, is for flexibility itself.

Yet all animals inhabit the same uncertain world that we do. The experience of a wild mouse is unfamiliar to us—following the scent of urine trails, gnawing on features of the environment, feeling in the dark with whiskers—but its survival also requires flexibility. Weather is variable, new predators come to town unpredictably, and food sources appear and disappear. All mammals may take advantage of a flexible internet-like architecture for similar reasons as humans.

In general, I predict that animals with more flexible, sophisticated routing schemes will show more flexible, sophisticated behavior. Flexibility is a characteristic of most mammals, and especially of primates. As a result, I believe all mammals have flexible routing systems with error correction, like the internet.

Nonmammals may not possess or need the most sophisticated or specialized features of internet-like protocol. For example, I predict that reptiles (and other animals with less sophisticated brains) will have a shorter protocol "stack." As I argued in chapter 7, the stack could correspond to the layered (laminated) structure of the cortex. While all placental mammals have six layers of cells in their cortex, indicating a rather sophisticated "stack," reptile cortices have only three layers.[1]

In nonmammal brains (and perhaps also in some small mammals), the organization of information passing from single neurotransmitter molecules up to whole-brain oscillations may have fewer functional subdivisions. Routing over more than one or two hops may also be unnecessary. In small animals, it is more feasible to connect neurons directly with most others—a 20-kilometer-wide head is only needed to interconnect all neurons in humans because we have so many more neurons that need interconnecting.

Routing protocol may be simpler still in small invertebrates like the honeybee. Yet arthropods (bees, spiders, ants, shrimp) are extraordinarily clever, especially for their size. Fruit flies, for example, can be trained to visually differentiate geometric patterns, as shown in an experiment by neurobiologist Martin Heisenberg and colleagues. The flies, with only 100,000 neurons in their brains, could even generalize this learning to some extent.[2]

Arthropod brains also deal with many kinds of messages on the same network. Their nervous systems are organized as a set of distinct centers—one for vision, another for smell—along with a set of relay stations that interconnect sensory systems with motor systems. Despite lacking a brain stem and spinal cord, all arthropods have a core brain region called the central body and an associated complex of helper regions, together called the central complex. The job of the central complex is partly to match predicted motions of the body and the environment to the actual motions of the body and the environment. The central complex compares signals produced by the brain to generate motor output in the environment (motion) with sensory signals feeding back from the environment to the nervous system via eyes, legs, antennae, and the like.

An arthropod brain may not permit vertebrate-like sophistication in flexibility or awareness. Arthropod survival generally relies more on numbers than on sophisticated behavior; there are whole clades of vertebrates specialized for the eating of insects. But one thing we can say is that, in arthropod brains and in our brains, the internal language of vision must intermingle with the language of smell and of inertial movements of the body. With more research guided by the internet metaphor, perhaps we will see that invertebrate brains are more sophisticated after all, especially in the way they intercommunicate. Future research of this kind will be aided by the publication of the largest connectome to date, for the fruit fly, containing twenty-five thousand neurons and several million synapses.[3]

Ultimately, different types of information must intermingle in the brains of all animals. Animals probably adopt routing strategies that resemble the internet to a greater or lesser extent depending on their living circumstances.

Nor do humans necessarily have the most internet-like brains. Instead, our brains may have adopted certain internet-like strategies and supercharged them. In particular, our brains are especially flexible. It is the ability to flexibly manage lots of information—and incorporate new information—that gives us an edge.

Flexibility is our greatest asset as a species. We now have the ability to usurp natural selection and to decide which species survive and which do not, from microbes to megafauna to reefs. Our brain's system for message passing gives rise to our intentions about where we go next as a species. The internet inside our heads, which coordinates it all, is now given even more power over the Earth thanks in part to the internet *outside* our heads. Let's use this power wisely.

ACKNOWLEDGMENTS

I WAS introduced to a profound truth about brains while attending a scientific conference held, incongruously, at Hamburger University on the McDonald's corporate campus near Chicago. Many have noted this truth before me, and it was said in passing by a neuroscientist at the conference, as if it was totally obvious to everyone. Over a non-fast-food lunch, the neuroscientist mentioned that every human brain is utterly different from every other human brain. No two brains—even those of identical twins—have precisely the same internal structure.

This idea was easy enough to grasp, but I realized that its implications were far-reaching. Despite having different brains, individuals of our species are perfectly capable of interacting with and understanding one another and of performing the same tasks. We regularly feel, imagine, think, and behave in nearly identical ways. Just observe a religious ritual or the morning commute. The implication is that there are basic principles of brain function and organization that do not depend on the unique fine structure of a given brain. And there must be general rules that brains have to follow to possess species-wide capabilities of thinking and behavior.

I gazed away from lunch with the neuroscientist to walls hung with imitations of classical paintings that had burgers and fries cleverly inserted in the image. I reflected on the welcome news that brains must operate on common principles. As a physics graduate student at the time, it felt natural to me that basic rules would apply equally to all comparable systems. Two soccer balls may differ in terms of color, design, materials, and so on, but they will experience the same effects of gravity. Likewise, two electrons can only vary in a few ways, such as in energy and spin; they are otherwise indistinguishable. Determining these invariant properties—and ignoring irrelevant structure—was the key to theoretical understanding in physics. While variability in the fine structure of the brain is surely important, we stand little chance of understanding these small differences until we understand the most basic rules of brain organization, which should apply equally to all comparable brains. Starting from this insight, it has been my goal to start constructing a neuroscientific theory of brain organization from the point of view of intrabrain communication, using the internet as a guiding metaphor. I don't recall the name of the neuroscientist whose comment inspired me that day at Hamburger University, but I would like to thank him.

This book is dedicated to my partner Reanna Lavine, and to Wesley and Kestrel: thank you for all your love and support. I want to acknowledge my debt of gratitude to the late Jeffrey Greenspon, the best colleague, mentor, and friend anyone could ask for. Yan Hao deserves huge thanks for helping me form many ideas in this book. Others, including Marcos Nadal, Dan Rockmore, David Field, Barb Finlay, Marc-Thorsten Hütt, Chris McManus, Ao Kevin Tang, Thom Dunn, Peter Britton, Chris Redies, György Buzsáki, Dan Sheldon, and Lauren Pomerantz, made important contributions to various aspects of this book (but I alone am responsible for any errors). Finally, my deepest gratitude goes to Miranda Martin, my superb editor at Columbia University Press, along with her outstanding team.

NOTES

PREFACE

1. Dean Buonomano, *Brain Bugs* (New York: Norton, 2011), 189, 105.

2. Donald Hoffman, *The Case Against Reality: How Evolution Hid the Truth from Our Eyes* (New York: Norton, 2011), 123.

3. Gary Marcus, "Face It, Your Brain Is a Computer," *New York Times*, June 28, 2015, https://www.nytimes.com/2015/06/28/opinion/sunday/face-it-your-brain-is-a-computer.html.

4. Computational neuroscientist Konrad Kording has actually performed this kind of experiment for the purpose of showing how computer metaphor thinking misses important structure and behavior. He and an electrical engineer colleague applied standard tools of analysis from computer metaphor–driven neuroscience to an actual computer, namely, the microprocessor chip of the classic Atari game system. As a reporter for *Wired* magazine noted, "they couldn't find much other than the circuit's off switch." Anna Vlastis, "Tech Metaphors Are Holding Back Brain Research," *Wired*, June 22, 2017, https://www.wired.com/story/tech-metaphors-are-holding-back-brain-research/; Eric Jonas and Konrad Paul Kording, "Could a Neuroscientist Understand a Microprocessor?" *PLoS Computational Biology* 13, no. 1 (January 2017): e1005268.

5. For a standard model of computations in decision-making, see Joshua I. Gold and Michael N. Shadlen, "The Neural Basis of Decision Making," *Annual Review of Neuroscience* 30 (July 2007): 535–574.

1. THE INTERNET-BRAIN AND
THE COMPUTER-BRAIN

1. Tomaso Poggio, "Routing Thoughts" (Working Paper No. 258, MIT AI Laboratory, Cambridge, MA, 1984).

2. Another early instance of the computer network metaphor in relation to brain science came in 1995 from Daniel Wegner, who used the computer network image as a way to illuminate information sharing in social interaction. Daniel M. Wegner, "A Computer Network Model of Human Transactive Memory," *Social Cognition* 13, no. 3 (September 1995): 319–339.

3. John H. Reynolds and David J. Heeger, "The Normalization Model of Attention," *Neuron* 61, no. 2 (January 2009): 168–185.

4. David Marr, *Vision: A Computational Investigation into the Human Representation and Processing of Visual Information* (Cambridge, MA: MIT Press, 1982), 5.

5. For a review of this question, see Laurenz Wiskott, "How Does Our Visual System Achieve Shift and Size Invariance?" in *23 Problems in Systems Neuroscience*, ed. J. Leo van Hemmen and Terrence J. Sejnowski (New York: Oxford University Press, 2005), 322–340.

6. B. A. Olshausen, C. H. Anderson, and D. C. Van Essen, "A Neurobiological Model of Visual Attention and Invariant Pattern Recognition Based on Dynamic Routing of Information," *Journal of Neuroscience* 13, no. 11 (1993): 4700–4719. See also P. Wolfrum, "Switchyards-Routing Structures in the Brain," in *Information Routing, Correspondence Finding, and Object Recognition in the Brain* (Berlin: Springer, 2010), 69–89.

7. Barbara Finlay, personal communication, March 13, 2019.

8. On Facebook, users are typically four degrees (friends) away from each other. Lars Backstrom et al., "Four Degrees of Separation," in *Proceedings of the Fourth Annual ACM Web Science Conference* (New York: Association for Computing Machinery, 2012), 33–42.

9. See, for example, Olaf Sporns, *Discovering the Human Connectome* (Cambridge, MA: MIT Press, 2012).

10. Georg F. Striedter, *Principles of Brain Evolution* (Sunderland, MA: Sinauer Associates, 2005), 126–131.

11. Thomas Weiss et al., "Rapid Functional Plasticity of the Somatosensory Cortex After Finger Amputation," *Experimental Brain Research* 134, no. 2 (2000): 199–203.

12. Philip Lieberman, *The Unpredictable Species: What Makes Humans Unique* (New York: Princeton University Press, 2013), 2.

13. Saak V. Ovsepian, "The Dark Matter of the Brain," *Brain Structure and Function* 224, no. 3 (2019): 973–983.

14. Michel A. Hofman, "Energy Metabolism, Brain Size and Longevity in Mammals," *Quarterly Review of Biology* 58, no. 4 (1983): 495–512.

15. Suzana Herculano-Houzel, "Scaling of Brain Metabolism with a Fixed Energy Budget per Neuron: Implications for Neuronal Activity, Plasticity and Evolution," *PLoS One* 6, no. 3 (2011).

16. Two conversations at once were possible on party lines, where everyone in the same building or block shared a single line. But intercommunication was never reliable in such systems, and it almost inevitably led to comedy and tears.

17. S. R. Cajal, *Leyes de la Morfología y Dinamismo de las Células Nerviosas* (Madrid, Spain: Imprenta y Librería de Nicolás Moya, 1897).

18. Barbara L. Finlay, "Principles of Network Architecture Emerging from Comparisons of the Cerebral Cortex in Large and Small Brains," *PLoS Biology* 14, no. 9 (2016): e1002556.

19. While only 5 percent of visual system inputs to the thalamus are from the retina, these fibers drive thalamic neurons to a greater extent than fibers of cortical origin. Still, the effective input from the retina to the thalamus is reckoned to be 10 percent or less of the total input. Masoud Ghodrati, Seyed-Mahdi Khaligh-Razavi, and Sidney R. Lehky, "Towards Building a More Complex View of the Lateral Geniculate Nucleus: Recent Advances in Understanding Its Role," *Progress in Neurobiology* 156 (2017): 214–255.

20. Farran Briggs and W. Martin Usrey, "A Fast, Reciprocal Pathway Between the Lateral Geniculate Nucleus and Visual Cortex in the Macaque Monkey," *Journal of Neuroscience* 27, no. 20 (2007): 5431–5436.

21. Anthony J. Bell, "Levels and Loops: The Future of Artificial Intelligence and Neuroscience," *Philosophical Transactions of the Royal Society of London, Series B: Biological Sciences* 354, no. 1392 (1999): 2013–2020.

22. Donald Hebb, *The Organization of Behavior: a Neuropsychological Theory* (New York: Wiley, 1949, repr. Lawrence Erlbaum Associates, 2002).

23. Jerzy Konorski, *Conditioned Reflexes and Neuron Organization* (Cambridge: Cambridge University Press, 1948), 89; Hebb, *The Organization of Behavior.*

24. For current views on Hebbian learning, see Patrick C. Trettenbrein, "The Demise of the Synapse as the Locus of Memory: A Looming Paradigm Shift?" *Frontiers in Systems Neuroscience* 10 (2016): 88.

25. See, for example, Alexander Schlegel, Prescott Alexander, and Peter U. Tse, "Information Processing in the Mental Workspace Is Fundamentally Distributed," *Journal of Cognitive Neuroscience* 28, no. 2 (2016): 295–307.

26. Pascal Fries, "A Mechanism for Cognitive Dynamics: Neuronal Communication Through Neuronal Coherence," *Trends in Cognitive Sciences* 9, no. 10 (2005): 474–480.

2. METAPHORS FOR THE BRAIN

1. Electron spin is quantized, unlike the angular momentum of a spinning bar magnet. But the conception of spin for an electron is otherwise identical to that of a spinning bar magnet.

2. Charles Lowney, "Rethinking the Machine Metaphor Since Descartes: On the Irreducibility of Bodies, Minds, and Meanings," *Bulletin of Science, Technology & Society* 31, no. 3 (2011): 179–192.

3. Steve Horvath and Kenneth Raj, "DNA Methylation-Based Biomarkers and the Epigenetic Clock Theory of Ageing," *Nature Reviews Genetics* 19, no. 6 (2018): 371.

4. Paul Cisek, "Beyond the Computer Metaphor: Behaviour as Interaction," *Journal of Consciousness Studies* 6, nos. 11–12 (1999): 125–142.

5. Sophisticated plumbing—including flush toilets supplied with water by terracotta pipes—has been found in the 3,500-year-old palace of Minos in Crete. Margalit Fox, *Riddle of the Labyrinth* (New York: HarperCollins, 2013), 32.

6. U.S. Census, "Historical Census of Housing Tables: Plumbing Facilities," https://www.census.gov/hhes/www/housing/census/historic/plumbing.html.

7. Engineering and Technology History Wiki, "Versailles Fountains," https://ethw.org/Versailles_Fountains.

8. René Descartes, *Treatise of Man*, trans. P. R. Sloan, in *The History and Philosophy of Science*, ed. Daniel McKaughan and Holly VandeWall (London: Bloomsbury Academic, 2018), 706.

9. Motoy Kuno, *The Synapse* (New York: Oxford University Press, 1995), 3.

10. Elliott S. Valenstein, *The War of the Soups and the Sparks* (New York: Columbia University Press, 2007), 3.

11. Peter Sterling and Simon Laughlin, *Principles of Neural Design* (Cambridge, MA: MIT Press, 2015), 106.

12. It should be noted that the camera obscura was discovered six hundred years earlier and analogized with the eye by the great scientist Ibn al-Haytham in Baghdad.

13. David Hockney, *Secret Knowledge: Rediscovering the Lost Techniques of the Old Masters* (New York: Viking, 2001).

14. This belief has roots in Plato and other classical thinkers, and it was first refuted by al-Haytham. But extramission theory may reflect a deeper folk psychology about awareness and knowledge. We know we are separate from the world but yet we can discover rich information from the stuff in it—what the stuff is and what it can do. Our awareness of this knowledge seemingly occurs instantaneously. Since we can't see the inner working of our brains, a natural hypothesis is that our eyes "ask" the world what is what, and the essence of each object provides a response. This is a natural deduction, and one more prevalent in children. By one measure, half of a sample of U.S. college students endorsed extramission theory. Gerald A. Winer et al., "Fundamentally Misunderstanding Visual Perception: Adults' Belief in Visual Emissions," *American Psychologist* 57, nos. 6–7 (2002): 417. The actual prevalence of extramission theory is probably lower, but it seems likely that extramission is humans' default assumption about how vision works.

15. Thomas Hobbes, *Leviathan* (London: Andrew Crooke, 1651), 2.

16. David Quammen, *The Tangled Tree* (New York: Simon and Schuster, 2018), 33.

17. Quoted in Howard E. Gruber and Paul H. Barrett, *Darwin on Man: A Psychological Study of Scientific Creativity* (London: Wildwood House, 1974), 451.

18. Danielle S. Bassett and Michael S. Gazzaniga, "Understanding complexity in the human brain," *Trends in Cognitive Sciences* 15, no. 5 (2011): 204.

19. Philip W. Anderson, "More Is Different," *Science* 177, no. 4047 (1972): 393–396.

20. Patricia S. Churchland and Terrence J. Sejnowski, *The Computational Brain* (Cambridge, Ma: MIT press, 1992), 7.

21. Flo Conway and Jim Siegelman, *Dark Hero of the Information Age: In Search of Norbert Wiener, the Father of Cybernetics* (New York: Basic Books, 2006).

22. Warren S. McCulloch and Walter Pitts, "A Logical Calculus of the Ideas Immanent in Nervous Activity," *Bulletin of Mathematical Biophysics* 5, no. 4 (1943): 115–133. The title of this paper harkens back to Leibniz, who coined the phrase "logical calculus."

23. McCulloch and Pitts, "A Logical Calculus," 115.

24. Warren S. McCulloch, "The Brain as a Computing Machine," *Electrical Engineering* 68 (1949): 492–497.

25. Frank Rosenblatt, "The Perceptron: a Probabilistic Model for Information Storage and Organization in the Brain." *Psychological Review* 65, no. 6 (1958): 386.

26. Although early deep learning researchers made use of key ideas about visual processing adapted from models of the mammal visual system, such as the Neocognitron artificial neural network of Kunihiko Fukushima, the concerted use of deep learning to study the visual brain began only in the twenty-first century. Kunihiko Fukushima, "Neocognitron: A Hierarchical Neural Network Capable of Visual Pattern Recognition," *Neural Networks* 1, no. 2 (January 1988): 119–130.

27. The number of weights grows at an accelerating rate as more units (neurons) are added to the deep net. The computing cost of manipulating arrays of weight values likewise grows in accelerating fashion. Processing speed limited these advances until the early 2000s. Today, deep net computations are performed on supercharged graphics cards that are specialized for manipulating large matrices of numbers (normally in the form of images).

28. Sometimes nodes also get information via connections to all the other nodes in the same layer. This is termed a Hopfield network.

29. Weight values representing a massive training set are the main store of value in a deep net, rather than any particular feature of the instructions in the computer code. However, weight values can be used to reverse-engineer the training data. In the case of cars and airplanes, this is harmless. But for faces and other identifying information, it is a major vulnerability. This is one of the reasons that deep learning AI researchers are increasingly keeping their programs secret: sensitive data like our daily habits or photos, scraped from the internet, can

be reconstructed from the weights of a deep net trained on our data. Madhumita Murgia, "Why Some AI Research May Be Too Dangerous to Share," *Financial Times*, June 19, 2019, https://www.ft.com /content/131f0430-9159-11e9-b7ea-60e35ef678d2.

30. See, for example, E. Kussul et al., "Rosenblatt Perceptrons for Handwritten Digit Recognition," in *IJCNN'01, International Joint Conference on Neural Networks Proceedings* (New York: Institute of Electrical and Electronics Engineers, 2001), 2: 1516–1520.

31. Xinge Zhu et al., "Dependency Exploitation: A Unified CNN-RNN Approach for Visual Emotion Recognition," in *Proceedings of the Twenty-Sixth International Joint Conference on Artificial Intelligence* (International Joint Conferences on Artificial Intelligence, 2017), 3595–3601.

32. Daniel L. K. Yamins et al., "Performance-Optimized Hierarchical Models Predict Neural Responses in Higher Visual Cortex," *Proceedings of the National Academy of Sciences* 111, no. 23 (2014): 8619–8624.

33. Approaches that attempt to predict blood flow–related measurements (i.e., fMRI signals that are distant echoes of spikes) in the visual system using deep learning models trained on natural images fare about the same in terms of explained variance. Radoslaw Martin Cichy et al., "Comparison of Deep Neural Networks to Spatio-Temporal Cortical Dynamics of Human Visual Object Recognition Reveals Hierarchical Correspondence," *Scientific Reports* 6 (2016): 27755.

34. Yamins et al. "Performance-Optimized Hierarchical Models Predict Neural Responses," *Proceedings of the National Academy of Sciences* 111, no. 23 (2014): 8619–8624.

35. Others have argued about the risks of relying on deep learning in AI as well; see, for example, Brenden M. Lake et al., "Building Machines That Learn and Think like People," *Behavioral and Brain Sciences* 40 (2017); and Gary Marcus, "Deep Learning: A Critical Appraisal," *arXiv: 1801 00631* (2018).

36. Le Chang and Doris Y. Tsao, "The Code for Facial Identity in the Primate Brain," *Cell* 169, no. 6 (2017): 1013–1028.

37. For example, in network science (a field comprising physicists, mathematicians, social scientists, and many others), the standard textbook, *Networks: An Introduction*, doesn't mention artificial neural networks at all. M. E. J. Newman, *Networks: An Introduction* (Oxford: Oxford University Press, 2010).

38. Mark E. Nelson and James M. Bower, "Brain Maps and Parallel Computers," *Trends in Neurosciences* 13, no. 10 (1990): 403–408.

39. Jonathan B. Levitt and Jennifer S. Lund, "Intrinsic Connections in Mammalian Cerebral Cortex," in *Cortical Areas: Unity and Diversity*, ed. A. Schüz and R. Miller (Boca Raton, FL: CRC Press, 2002), 145–166.

40. Michèle Fabre-Thorpe et al., "A Limit to the Speed of Processing in Ultra-Rapid Visual Categorization of Novel Natural Scenes," *Journal of Cognitive Neuroscience* 13, no. 2 (2001): 171–180.

41. Martin J. Tovée, "Neuronal Processing: How Fast Is the Speed of Thought?" *Current Biology* 4, no. 12 (1994): 1125–1127.

42. See, for example, Cathy O'Neill, *Weapons of Math Destruction: How Big Data Increases Inequality and Threatens Democracy* (New York: Broadway Books, 2017).

43. Seyed-Mahdi Khaligh-Razavi and Nikolaus Kriegeskorte, "Deep Supervised, but Not Unsupervised, Models May Explain IT Cortical Representation," *PLoS Computational Biology* 10, no. 11 (2014).

44. We can see the invariance problem in action using an excellent online visualization of a deep net designed for visual recognition by Ryerson University researchers: "An Interactive Node-Link Visualization of Convolutional Neural Networks," http://scs.ryerson.ca/~aharley/vis/conv. The simulator tries to identify the digits 0–9 when we write them on a screen. It displays the activity of each neuron in each layer of the network as a color. Try writing a few numbers—it works pretty well! Now try writing them elongated, and only in half the frame. The system's guesses will be wrong almost every time. Distortions that are easy for humans to ignore in identifying images can render deep networks totally ineffective. Today, sophisticated deep nets fare better at this task compared to this demonstration, but the basic problem of invariant object recognition in natural images remains unsolved.

45. Frederic W. H. Myers, "Multiplex personality," *Proceedings of the Society for Psychical Research*, 4 (1886-87), 503.

46. Charles S. Sherrington, *Man on His Nature.* (New York: Macmillan, 1941), 225.

47. Leibniz himself was mistrustful of the mill metaphor, and in fact proposed it in the context of suggesting that subjective experiences such as visual perception could *not* be conceived as mechanical processes. G. W. Leibniz, "The Principles of Philosophy, or, the Monadology," in

Philosophical Essays, ed. and trans. R. Ariew and D. Garber (Indianapolis, IN: Hackett, 1989), 213–225. (Original work published 1714.)

48. Herbert Spencer, *Principles of Psychology* (New York: D. Appleton, 1896), 529.

49. Stephen Jay Gould, *The Structure of Evolutionary Theory* (Cambridge, MA: Harvard University Press, 2002), 197.

50. Ivan P. Pavlov, *Conditioned Reflexes: An Investigation of the Physiological Activity of the Cerebral Cortex*, trans. G. V. Anrep (Oxford: Oxford University Press, 1927), 25–26; 219.

51. Charles Sherrington, *Integrative Action of the Nervous System* (Cambridge: Cambridge University Press, 1947), 234.

52. Claude E. Shannon, "The Bandwagon," *IRE Transactions on Information Theory* 2, no. 1 (1956): 3.

53. Andrea Goldsmith et al., "Beyond Shannon: The Quest for Fundamental Performance Limits of Wireless ad Hoc Networks," *IEEE Communications Magazine* 49, no. 5 (2011): 195–205. See also Abbas El Gamal and Young-Han Kim, *Network Information Theory* (Cambridge: Cambridge University Press, 2011). Another promising general strategy is collectively known as graph-based entropy, though this approach cannot at present account for communication dynamics on complex networks. Matthias Dehmer and Abbe Mowshowitz, "A History of Graph Entropy Measures," *Information Sciences* 181, no. 1 (2011): 57–78.

3. WHAT WE DON'T KNOW ABOUT BRAINS

1. Frederico A. C. Azevedo et al., "Equal Numbers of Neuronal and Nonneuronal Cells Make the Human Brain an Isometrically Scaled-Up Primate Brain," *Journal of Comparative Neurology* 513, no. 5 (2009): 532–541.

2. Bosiljka Tasic et al., "Shared and Distinct Transcriptomic Cell Types Across Neocortical Areas," *Nature* 563, no. 7729 (2018): 72.

3. See Horace Basil Barlow, "Why Have Multiple Cortical Areas?" *Vision Research* 26, no. 1 (1986): 81–90. The challenge in measuring how many neurons are connected to a typical neuron is that it is very hard to get a representative sample across the whole brain. We might imagine we could measure connections in a few neurons in a given area and then

extrapolate to larger areas and ultimately to the whole brain. A related approach was tried for many years by anatomists trying to estimate the simple number of neurons in the human brain. They came up with widely varying estimates. The solution to this problem, employed by Suzana Herculano-Houzel and colleagues, was to literally put brains in a blender with chemicals that dissolved neurons' cell membranes. Examining the resulting slurry under the microscope, Herculano-Houzel counted neuron cell nuclei. Leveraging the fact that there is one nucleus per neuron, she arrived at the most accurate count of neurons to date. See Suzana Herculano-Houzel, *The Human Advantage: A New Understanding of How our Brain Became Remarkable* (Cambridge, MA: MIT Press, 2016). But the blender approach would not work for estimating connection numbers because they would be destroyed in the process. Consequently, typical connection numbers remain unknown.

4. Gabriel Kreiman, "Neural Coding: Computational and Biophysical Perspectives," *Physics of Life Reviews* 1, no. 2 (2004): 71–102.

5. A. Paul Alivisatos et al., "The Brain Activity Map Project and the Challenge of Functional Connectomics," *Neuron* 74, no. 6 (2012): 970–974.

6. There are several estimates of energy cost, computed in various ways. See William B. Levy and Robert A. Baxter, "Energy Efficient Neural Codes," *Neural Computation* 8 (1997): 531–543; David Attwell and Simon B. Laughlin, "An Energy Budget for Signaling in the Grey Matter of the Brain," *Journal of Cerebral Blood Flow & Metabolism* 21, no. 10 (2001): 1133–1145; and Peter Lennie, "The Cost of Cortical Computation," *Current Biology* 13, no. 6 (2003): 493–497. All estimates suggest that the actual proportion of the population that is highly active may be even lower than 10 percent.

7. But perhaps the study of cell signaling is also due for a fundamental rethink of the role of routing. Cell biologist John Hancock, author of a textbook on cell signaling, asks, "If both hormone X and hormone Y lead to an increase in the concentration of signal Z inside the cell, when signal Z increases, how does the cell know that it is due to the perception of X or Y?" Hancock believes that it is due to emergent complexity. John T. Hancock, *Cell Signalling* (Oxford: Oxford University Press, 2017), 330.

8. Edward M. Callaway and Anupam K. Garg, "Brain Technology: Neurons Recorded en Masse," *Nature* 551, no. 7679 (2017): 172.

9. John Carew Eccles, *The Physiology of Synapses* (New York: Academic Press, 1964), 1–2.

10. Wulfram Gerstner and Richard Naud, "How Good Are Neuron Models?" *Science* 326, no. 5951 (2009): 379–380.

11. Michael Häusser and Bartlett Mel, "Dendrites: Bug or Feature?" *Current Opinion in Neurobiology* 13, no. 3 (2003): 372–383.

12. Jérôme Epsztein et al., "Impact of Spikelets on Hippocampal CA1 Pyramidal Cell Activity During Spatial Exploration," *Science* 327, no. 5964 (2010): 474–477. When signals go the wrong way in a neuron, there is a distinctive sensation, at least in neurons of the somatosensory system. This is the funny bone tingle. See Gerald E. Schneider, *Brain Structure and Its Origins: In Development and in Evolution of Behavior and the Mind* (Cambridge, MA: MIT Press, 2014), 17.

13. David H. Hubel and Torsten N. Wiesel, "Receptive Fields, Binocular Interaction and Functional Architecture in the Cat's Visual Cortex," *Journal of Physiology* 160, no. 1 (1962): 106–154.

14. David H. Hubel and Torsten N. Wiesel, "Receptive Fields of Single Neurones in the Cat's Striate Cortex," *Journal of Physiology* 148, no. 3 (1959): 574–591.

15. Today, with magnetic resonance imaging, we can watch the activity of neurons across a stretch of the cortical surface of V1 and see that it roughly corresponds to the bright parts of the retinal image. Reading out retinotopy is the dominant strategy in brain imaging studies that aim to perform "brain reading." These studies can achieve quite high accuracy in guessing which image out of a set of a thousand a person is looking at solely by examining their brains in the imaging scanner. Kendrick N. Kay et al., "Identifying Natural Images from Human Brain Activity," *Nature* 452, no. 7185 (2008): 352.

16. Hubel and Wiesel proposed a similar wiring model explaining how simple cells could pool together inputs from the retina to make oriented rectangle detectors. Hubel and Wiesel, "Receptive Fields, Binocular Interaction."

17. Hubel and Wiesel's wiring model is plausible, but it remains controversial. The wiring model for inputs to simple cells is supported by indirect evidence from neural recordings (see Nicholas J. Priebe, "Mechanisms of orientation selectivity in the primary visual cortex," *Annual Review of Vision Science* 2 (2016): 85–107) but the complex cell wiring model

has little empirical support. There is even still debate about whether there is a meaningful distinction between simple and complex cells, or if instead V1 neurons form more of a continuum of preferences from simple-like cells to complex-like cells. Larry F. Abbott and Frances S. Chance, "Rethinking the Taxonomy of Visual Neurons," *Nature Neuroscience* 5, no. 5 (2002): 391; Hubel and Wiesel, "Receptive Fields, Binocular Interaction."

18. Interestingly, Hubel and Wiesel did stain and examine the brain tissue they recorded from (after sacrificing the animal), but not in a way that permitted them to trace the connections of the cells they recorded from. Simple and complex cells differ in function, but they cannot be distinguished based on morphology (i.e., how they look under the microscope after staining). Hubel and Wiesel, "Receptive Fields, Binocular Interaction."

19. There is a basic limitation in recording neural activity that is a bit reminiscent of the Heisenberg Uncertainty Principle: one generally cannot measure a neuron's anatomy—such as its connectivity to other neurons—while also measuring its physiology (its electrical responses, for example).

20. Austin E. Soplata et al., "Thalamocortical Control of Propofol Phase-Amplitude Coupling," *PLOS Computational Biology* 13, no. 12 (2017): e1005879.

21. Horace B. Barlow, "Possible Principles Underlying the Transformation of Sensory Messages," in *Sensory Communication*, ed W. A. Rosenblith, 1–34 (Cambridge, MA: MIT Press, 1961). Barlow cited earlier work by Fred Attneave at the University of Oregon and others going back to the polymath physicist Ernst Mach in the nineteenth century. A parallel school of thought emerged in experimental biology slightly later. Starting in the late 1970s, John Lythgoe, working with marine animals, focused on the particular challenges of vision in murky or dark or turbulent environments and the kinds of solutions needed in different parts of the brain. John Nicholas Lythgoe, *Ecology of Vision* (London: Clarendon Press, 1979).

22. Barbara L. Finlay, Luiz Carlos de Lima Silveira, and Andreas Reichenbach, "Comparative Aspects of Visual System Development," in *The Primate Visual System: A Comparative Approach*, ed. Jan Kremers (Chichester, UK: John Wiley, 2005), 37–72.

23. Retinal ganglion cell neurons turn out to have considerable correlations in firing with their neighbors. Marsha Meytlis, Zachary Nichols, and Sheila Nirenberg, "Determining the Role of Correlated Firing in Large Populations of Neurons Using White Noise and Natural Scene Stimuli," *Vision Research* 70 (2012): 44–53. It is also debated whether retaining some correlations—that is, some redundancy—is actually beneficial for the uses brains make of sense data. We will return to this question in chapter 8.

24. Michael S. Lewicki, "Efficient Coding of Natural Sounds," *Nature Neuroscience* 5, no. 4 (2002): 356.

25. Karl Friston, "The Free-Energy Principle: A Rough Guide to the Brain?" *Trends in Cognitive Sciences* 13, no. 7 (2009): 293–301.

26. Simon Laughlin, "A Simple Coding Procedure Enhances a Neuron's Information Capacity," *Zeitschrift für Naturforschung C* 36, nos. 9–10 (1981): 910–912.

27. Both sets of researchers simply measured the similarity of all pixels and their neighbors using high-resolution photographs of the world, making sure that pixels in the digitized images corresponded to actual light intensities in each location (cameras normally squeeze high-intensity pixels into lower intensities but leave darker pixels mostly as they are; thus calibration was required). Measurements of this kind had been taken by early television engineers—even before Barlow's seminal paper—but apparently these results were unknown to Barlow. The universality of the law of similar neighboring pixels—and the fact that it holds regardless of the physical size of the pixels, giving it scale invariance—was uncovered both by Field and Burton and Moorhead. These researchers had the benefit of digitized images and powerful desktop computers, which Barlow lacked, to perform measurements of correlations in pixel values. David J. Field, "Relations Between the Statistics of Natural Images and the Response Properties of Cortical Cells," *Journal of the Optical Society of America A* 4, no. 12 (1987): 2379–2394; Geoffrey J. Burton and Ian R. Moorhead, "Color and Spatial Structure in Natural Scenes," *Applied Optics* 26, no. 1 (1987): 157–170.

28. There are also written languages that split the difference. Scripts such as Linear B, a form of proto-Greek from the second century B.C.E., had around eighty symbols that represented syllables—*wa* or *po*—rather than the phonemes like *w*- and *p*-, which need to be combined with a

vowel in written English. Linear B also included pictographs similar to those used in Chinese.

29. In practice, Chinese is not as sparse as the ideal case, since some symbols are used very frequently, such as those that mean *one* and *yes*.

30. Sparseness is not a measure of total or average activity; it is what's called a higher-order statistic. A set of neurons with a given level of total activity could have a few units highly active and the rest off, or it could have most units partially active. Statistically, sparseness measures how peaky and heavy-tailed the distribution of activities is. A peak near zero would indicate that most units are off, while heavy tails indicate that some are highly active. It is often gauged using the kurtosis (the fourth statistical moment of a distribution, coming after mean, variance, and skew) of the distribution. See Benjamin Willmore and David J. Tolhurst, "Characterizing the Sparseness of Neural Codes," *Network: Computation in Neural Systems* 12, no. 3 (2001): 255–270; and Daniel J. Graham and David J. Field, "Sparse Coding in the Neocortex," in *Evolution of Nervous Systems*, ed. Jon H. Kaas and Leah A. Krubitzer (Amsterdam: Academic Press, 2007), 3: 181–187.

31. Daniele Poli et al., "Sparse and Specific Coding During Information Transmission Between Co-cultured Dentate Gyrus and CA3 Hippocampal Networks," *Frontiers in Neural Circuits* 11 (2017): 13. See also Bruno A. Olshausen and David J. Field, "What Is the Other 85 Percent of V1 Doing?" in *23 Problems in Systems Neuroscience*, ed. J. Leo van Hemmen and Terrence J. Sejnowski (New York: Oxford University Press, 2005), 182–211.

32. Bruno A. Olshausen and David J. Field, "Emergence of Simple-Cell Receptive Field Properties by Learning a Sparse Code for Natural Images," *Nature* 381, no. 6583 (1996): 607–609.

33. Concurrent work by computational neuroscientists Tony Bell and Terry Sejnowski reached the same conclusion and provided a more generalized mathematical framework for sparse coding. Anthony J. Bell and Terrence J. Sejnowski, "The 'Independent Components' of Natural Scenes Are Edge Filters," *Vision Research* 37, no. 23 (1997): 3327–3338. For an overview, see Aapo Hyvärinen, Jarmo Hurri, and Patrick O. Hoyer, *Natural Image Statistics: A Probabilistic Approach to Early Computational Vision* (Berlin: Springer Science and Business Media, 2009).

34. William E. Vinje and Jack L. Gallant, "Sparse Coding and Decorrelation in Primary Visual Cortex During Natural Vision," *Science* 287, no. 5456 (2000): 1273–1276.

35. Shy Shoham, Daniel H. O'Connor, and Ronen Segev, "How Silent Is the Brain: Is There a 'Dark Matter' Problem in Neuroscience?" *Journal of Comparative Physiology A* 192, no. 8 (2006): 777–784.

36. One idea of how the brain came to have so many dormant neurons is that they are actively inhibited by other circuits. Being prevented from firing via neurotransmitters from active neurons could mean that the dark neurons can no longer be weeded out by evolution. See J. Lee Kavanau, "Conservative Behavioural Evolution, the Neural Substrate," *Animal Behaviour* 39, no. 4 (1990): 758–767.

37. Erik Learned-Miller et al., "Labeled Faces in the Wild: A Survey," in *Advances in Face Detection and Facial Image Analysis*, ed. M. Kawulok et al. (Cham, Switzerland: Springer, 2016), 189–248.

38. Wiktor Młynarski and Josh H. McDermott, "Ecological Origins of Perceptual Grouping Principles in the Auditory System," *Proceedings of the National Academy of Sciences* 116, no. 50 (2019): 25355–25364.

39. Field, who was my Ph.D. advisor, has taken the principle of sparse coding to heart, or to brain as it were. Given the limit of 10 percent or fewer cells highly active at a time in the brain, David likes to joke that he is trying to get his personal total down to 5 percent. He may be on to something: there is now evidence (albeit shaky) showing that lower activity in the cortex leads to longer life spans: Joseph M. Zullo et al., "Regulation of Lifespan by Neural Excitation and REST," *Nature* 574, no. 7778 (2019): 359–364.

4. FROM CONNECTOMICS TO DYNOMICS

1. Olaf Sporns, Giulio Tononi, and Rolf Kötter, "The Human Connectome: A Structural Description of the Human Brain," *PLoS Computational Biology* 1, no. 4 (2005). The same term was used in the same way in the same year in the following Ph.D. thesis: Patric Hagmann, "From Diffusion MRI to Brain Connectomics" (Ph.D. diss., Swiss Federal Institute of Technology Lausanne, 2005), 108.

2. For a lively description of this technology, see Sebastian Seung, *Connectome: How the Brain's Wiring Makes Us Who We Are* (New York: Houghton Mifflin Harcourt, 2012).

3. Mark D. Humphries, Kevin Gurney, and Tony J. Prescott, "The Brain-stem Reticular Formation Is a Small-World, Not Scale-Free, Network," *Proceedings of the Royal Society B: Biological Sciences* 273, no. 1585 (2006): 503–511.

4. Laurenz Wiskott and Christof von der Malsburg, "Face Recognition by Dynamic Link Matching," in *Lateral Interactions in Cortex: Structure and Function*, ed. J. Sirosh, R. Miikkulainen, and Y. Choe (Austin, TX: UTCS Neural Networks Research Group, 1996), 1.

5. Martijn P. Van Den Heuvel and Olaf Sporns, "Rich-Club Organization of the Human Connectome," *Journal of Neuroscience* 31, no. 44 (2011): 15775–15786.

6. Răzvan Gămănuţ et al., "The Mouse Cortical Connectome, Characterized by an Ultra-Dense Cortical Graph, Maintains Specificity by Distinct Connectivity Profiles," *Neuron* 97, no. 3 (2018): 698–715.

7. Van Essen, working with Bruno Olshausen, whom we met in the previous chapter, and computational neuroscientist Charles H. Anderson, was the first to propose a detailed model of flexible "routing" in the brain. This work is the basis of the invariance mechanism for the visual system described in chapter 1, whereby proto-objects found anywhere in retinal space are brought into register with a single set of object detectors. This model was rather heretical when first proposed in 1993 and has mostly been ignored by the computer metaphor–driven field. But its salience today is becoming more widely recognized. Bruno A. Olshausen, Charles H. Anderson, and David C. Van Essen, "A Neuro-biological Model of Visual Attention and Invariant Pattern Recognition Based on Dynamic Routing of Information," *Journal of Neuroscience* 13, no. 11 (1993): 4700–4719.

8. Daniel J. Felleman and David C. Van Essen, "Distributed Hierarchical Processing in the Primate Cerebral Cortex," *Cerebral Cortex* 1, no. 1 (1991): 30.

9. Răzvan Gămănuţ et al., "The Mouse Cortical Connectome, Characterized by an Ultra-Dense Cortical Graph, Maintains Specificity by Distinct Connectivity Profiles," *Neuron* 97, no. 3 (2018): 698–715.

10. In addition, the mouse's primary visual cortex—a region traditionally seen as processing "only" signals arising from the eyes—seems to do other jobs. It has been found to fire reliably in total darkness during particular motor tasks. Georg B. Keller, Tobias Bonhoeffer, and Mark

Hübener, "Sensorimotor Mismatch Signals in Primary Visual Cortex of the Behaving Mouse," *Neuron* 74, no. 5 (2012): 809–815.

11. Nikola T. Markov et al., "Weight Consistency Specifies Regularities of Macaque Cortical Networks," *Cerebral Cortex* 21, no. 6 (2011): 1254–1272; Nikola T. Markov et al., "A Weighted and Directed Interareal Connectivity Matrix for Macaque Cerebral Cortex," *Cerebral Cortex* 24, no. 1 (2014): 17–36;

12. Danielle S. Bassett and Edward T. Bullmore, "Small-World Brain Networks Revisited," *The Neuroscientist* 23, no. 5 (2017): 499–516.

13. Tom Binzegger, Rodney J. Douglas, and Kevan A. C. Martin, "Topology and Dynamics of the Canonical Circuit of Cat V1," *Neural Networks* 22, no. 8 (2009): 1071–1078. There is debate about whether it is accurate to assume values for certain anatomical parameters in the work of Binzegger and colleagues, as well as others in this field. These parameters partially determine the researchers' estimates of connection strength (i.e., the number of synapses between cells). Others have argued that the fine structure of synaptic connection is important and must be measured directly (which is laborious and difficult). For further reading on this debate, see Olaf Sporns, *Discovering the Human Connectome* (Cambridge, MA: MIT Press, 2012), 75.

14. Mária Ercsey-Ravasz et al., "A Predictive Network Model of Cerebral Cortical Connectivity Based on a Distance Rule," *Neuron* 80, no. 1 (2013): 184–197; Szabolcs Horvát et al., "Spatial Embedding and Wiring Cost Constrain the Functional Layout of the Cortical Network of Rodents and Primates," *PLoS Biology* 14, no. 7 (2016): e1002512.

15. Again, V2 is an area assumed by classical computational models as only receiving important inputs from V1, rather than sending important inputs to V1.

16. Seung Wook Oh et al., "A Mesoscale Connectome of the Mouse Brain," *Nature* 508, no. 7495 (2014): 207.

17. While axons entering the thalamus from a given source area tend to terminate in the same layer, this is not its only characteristic. Gordon Walls, a pioneer in the comparative study of neuroanatomy and physiology in vertebrate visual systems, has described axons that pierce through layers of the thalamus like "toothpicks in a club sandwich," potentially delivering signals to multiple layers en route. This architecture could help transmit messages more widely via the thalamus. See:

Gordon Lynn Walls, "The Lateral Geniculate Nucleus and Visual His-
tophysiology," *University of California Publications in Physiology* 9, no. 1
(1953), 1; Masoud Ghodrati, Seyed-Mahdi Khaligh-Razavi, and Sidney
R. Lehky. "Towards Building a More Complex View of the Lateral
Geniculate Nucleus: Recent Advances in Understanding Its Role,"
Progress in Neurobiology 156 (2017): 217.

18. In humans, the anterior cingulate helps us monitor conflict, learn about
rewards, and perform social evaluation, among numerous other functions.

19. It is not necessarily the case that parallel fibers do not interact. It has
been known since 1940 that adjacent axons are entirely capable of
influencing each other, with one fiber potentially speeding up, slow-
ing down, or effecting synchronization of activity in other fibers. One
curious proposal is that interactions of parallel axon fibers underlie
the phenomenon of "photic sneeze." This is the condition where some
people reflexively sneeze when going from darkness into bright light.
Although evidence is limited, the idea is that a sudden burst of activity in the optic nerve (ganglion cell axons) due to light excites parallel
axons of the trigeminal nerve, which is involved in sneeze reflexes. Barrett Katz, Ronald B. Melles, Michael R. Swenson, and Jerry A. Schneider, "Photic Sneeze Reflex in Nephropathic Cystinosis," *British Journal
of Ophthalmology* 74, no. 12 (1990): 706–708. See also Hiba Sheheitli and
Viktor K. Jirsa. "A Mathematical Model of Ephaptic Interactions in
Neuronal Fiber Pathways: Could There be More than Transmission
Along the Tracts?" *Network Neuroscience* 4, no. 3 (2020): 595–610.

20. For a critique of this area, see Michael N. Hallquist and Frank G. Hillary, "Graph Theory Approaches to Functional Network Organization
in Brain Disorders: A Critique for a Brave New Small-World," *Network
Neuroscience* 3, no. 1 (2018): 1–26.

21. Nancy J. Kopell et al., "Beyond the Connectome: The Dynome," *Neuron*
83, no. 6 (2014): 1319.

5. HOW THE INTERNET WORKS

1. Romauldo Pastor-Satorras and Alessandro Vespignani, *Evolution and
Structure of the Internet: A Statistical Physics Approach* (Cambridge:
Cambridge University Press, 2009), 9. This book gives an exhaustive but
eminently readable examination of the physics of the internet (including mathematical descriptions of network topology and dynamics).

2. Leonard Kleinrock, *Queueing Systems, Vol. 2: Computer Applications* (New York: Wiley, 1976), vii.

3. Quoted in James Gillies and Robert Cailliau, *How the Web Was Born: The Story of the World Wide Web* (New York: Oxford University Press, 2000), 22. Much of the historical narrative of this chapter is drawn from Gillies and Calliau's book.

4. Sharla P. Boehm and Paul Baran, "On Distributed Communications: II. Digital Simulation of Hot-Potato Routing in a Broadband Distributed Communications Network," *Memorandum of the RAND corporation prepared for United States Air Force* (1964).

5. Gillies and Cailliau, *How the Web Was Born*, 25. The schemes of both Baran and Davies specified the same size of packets—1,024 bits—as well as nearly the same ordering of packet components, such as sender and receiver address.

6. As Davies showed, the channel also needs to be able to transmit more than 10,000 bits per second for packets to be effective—but this is a glacial rate by today's standards, and achieving it was not even a major challenge in the mid-1960s. James Pelkey, "Entrepreneurial Capitalism and Innovation: A History of Computer Communications 1968–1988," http://www.historyofcomputercommunications.info/Book/2/2.5%20Donald%20Davies-65-66.html. ARPANET ran at more than twice this speed. Barry M. Leiner et al., *A Brief History of the Internet*, Internet Society, 1997, https://www.internetsociety.org/internet/history-internet/brief-history-internet/.

7. Ethernet protocol generates random back-off times according to a decaying exponential distribution, meaning that the delay for a given channel is very likely to be brief, but the delay is still unlikely to be the same as that for another incoming channel (thereby preventing collision). We will discuss exponential back-off in greater detail in chapter 6.

8. See, for example, Marc-Thorsten Hütt et al., "Stochastic Resonance in Discrete Excitable Dynamics on Graphs," *Chaos, Solitons & Fractals* 45, no. 5 (2012): 611–618.

9. Paul Tilghman, "If DARPA Has Its Way, AI Will Rule the Wireless Spectrum," *IEEE Spectrum* 56, no. 6 (2019): 28–33.

10. Even when we know a packet has arrived successfully, there is a chance that it has been corrupted en route. Some of the ones and zeros that constitute the packet might have been changed, possibly rendering the whole packet—and thus the whole message—unintelligible. The

internet uses an additional scheme for identifying corrupted packets. As with acks, the solution is simple and has been employed successfully starting with ARPANET (though the basic solution predates the internet era by decades). The solution is called the *checksum*. Before a packet is sent, the ones and zeros in that message are added together, and the resulting sum is deposited in the packet. When a packet is delivered, the receiving machine adds together the ones and zeros again, and compares this value to the checksum. If the two values don't match, the receiver knows the packet was corrupted, and it tells the sender to resend the corrupted packet.

11. Newer internet protocol allows far more possible addresses than are available in the legacy protocol (IPv4). The new IPv6 protocol does this by making the numerical addresses much longer, but IPv6 has not been fully adopted.

12. I have described the internet's routing protocol in its ideal, theoretical form. The real internet of today is rather different. For one thing, the end of enforced net neutrality in the United States means that packets may not all be treated equivalently. Also, because even the slowest routers process packets within a few milliseconds, tiny packet-sized chunks of information are becoming less salient today. Instead, flows of packets are more relevant to minimizing congestion on the network. Flows bundle together many packets, such as those comprising a media stream, and can be subjected to their own routing rules.

13. Neurons spend most of their energy on moving ions in and out of the cell and on packaging up neurotransmitters in preparation for sending messages down the axon. David Attwell and S. B. Laughlin, "An Energy Budget for Signaling in the Grey Matter of the Brain," *Journal of Cerebral Blood Flow and Metabolism*, 21, no. 10 (2001): 1133–1145.

6. THE INTERNET METAPHOR: FIRST STEPS TO A NEW THEORY OF THE BRAIN

1. I have advocated for some of the ideas presented in this chapter in previous work: Daniel J. Graham and Daniel Rockmore, "The Packet Switching Brain," *Journal of Cognitive Neuroscience* 23, no. 2 (2011): 267–276; Daniel J. Graham, "Routing in the Brain," *Frontiers in Computational Neuroscience* 8 (2014): 44.

2. Karl Friston, "The Free-Energy Principle: A Unified Brain Theory?" *Nature Reviews Neuroscience* 11 (2010): 127–138.

3. The mathematics, it turns out, is mostly smokescreen. The issue is that working out the true mathematical likelihood of an event in the world—say, recognizing the identity of a face—is not usually achievable in practice. There are just too many things that can affect the probability of such an event. It's even harder to work out the likelihood of the neural encoding related to the event. Since the free-energy framework posits that the brain's overarching job is to minimize the difference between world likelihoods and neural encoding likelihoods, this mathematical formulation of the theory doesn't actually lead to implementations that can be tested in relation to real brains. As with deep nets, the Free-Energy Principle may be useful in engineering, especially in robotics, but this framework may not be as helpful in advancing our understanding of brains.

4. For background on the optimality debate in evolutionary biology, see, for example, Oscar Vilarroya, " 'Two' Many Optimalities," *Biology and Philosophy* 17, no. 2 (2002): 251–270.

5. Sabine Tebbich, Kim Sterelny, and Irmgard Teschke, "The Tale of the Finch: Adaptive Radiation and Behavioural Flexibility," *Philosophical Transactions of the Royal Society B: Biological Sciences* 365, no. 1543 (2010): 1099–1109.

6. A different theory holds that color vision evolved to help us make judgments about faces. The argument goes that among the placental mammals, only primates have full color vision. We are a uniquely social group that performs various degrees of "mind reading" based in part on face coloration. We have significant expanses of bare skin on our faces, so we notice immediately when someone appears "flushed." The improved color discrimination enabled by primate vision is well matched to these color differences. See Mark A. Changizi, Qiong Zhang, and Shinsuke Shimojo, "Bare Skin, Blood and the Evolution of Primate Colour Vision," *Biology Letters* 2, no. 2 (2006): 217–221; and Maryam Hasantash et al., "Paradoxical Impact of Memory on Color Appearance of Faces," *Nature Communications* 10, no. 1 (2019): 1–10.

7. Philosopher Daniel Milo argues that our search for optimality in the traits of living things stems from the erroneous application of the concept of artificial selection in the natural world. Whereas humans

can and do select plants and animals to optimally express some trait—from nutritional content in plants to cute puppy ears—nature is not so aggressive in its selection of single traits in complex animals. For example, single species show large variation in key traits. Our kidneys, a costly, indispensable organ, vary over an enormous range: the number of nephrons (subunits that filter blood) varies more than twelvefold. If kidneys were optimal, they would all be very similar. Instead, almost any number of nephrons within this range is good enough for most humans. Milo argues that we should therefore assume that traits are neutral until proven otherwise, rather than assuming that every trait is adaptive (and optimally so). Daniel S. Milo, *Good Enough* (Cambridge, MA: Harvard University Press, 2019).

8. A. Aldo Faisal, Luc P. J. Selen, and Daniel M. Wolpert, "Noise in the Nervous System," *Nature Reviews Neuroscience* 9, no. 4 (2008): 292.

9. David J. Heeger, "Poisson Model of Spike Generation," September 5, 2000, https://www.cns.nyu.edu/~david/handouts/poisson.pdf.

10. Before spiking again, there is a brief silent period, during which the cell must "recharge" its spiking machinery. This *refractory period* lasts a few milliseconds.

11. See Roland Baddeley et al., "Responses of Neurons in Primary and Inferior Temporal Visual Cortices to Natural Scenes," *Proceedings of the Royal Society of London B: Biological Sciences* 264, no. 1389 (1997): 1775–1783. See also Robbe L. T. Goris, J. Anthony Movshon, and Eero P. Simoncelli, "Partitioning Neuronal Variability," *Nature Neuroscience* 17, no. 6 (2014): 858; Alexander S. Ecker et al., "State Dependence of Noise Correlations in Macaque Primary Visual Cortex," *Neuron* 82, no. 1 (2014): 235–248; and Adam S. Charles et al., "Dethroning the Fano Factor: A Flexible, Model-Based Approach to Partitioning Neural Variability," *Neural Computation* 30, no. 4 (2018): 1012–1045.

12. See, for example, Rufin VanRullen, Rudy Guyonneau, and Simon J. Thorpe, "Spike Times Make Sense," *Trends in Neurosciences* 28, no. 1 (2005): 1–4.

13. Saskia E. J. de Vries, Jerome A. Lecoq, Michael A. Buice, Peter A. Groblewski, Gabriel K. Ocker, Michael Oliver, David Feng et al., "A Large-Scale Standardized Physiological Survey Reveals Functional Organization of the Mouse Visual Cortex," *Nature Neuroscience* 23, no. 1 (2020): 138.

14. For the sparseness of rat primary auditory cortex, see T. Hromadka, M. R. Deweese, and A. M. Zador, "Sparse Representation of Sounds in the Unanesthetized Auditory Cortex," *PLoS Biology* 6, no. 1 (2008): e16. See also Saak V. Ovsepian, "The Dark Matter of the Brain," *Brain Structure and Function* 224, no. 3 (2019): 973–983.

15. There are other forms of anatomical noise, including neurons in a given functional area that don't respond to any stimuli of the kind for which that area is specialized. For example, some cells in the primary visual cortex respond to sound and touch. There are also neurons in a given functional area that don't respond to the stimulus dimension we assume they prefer. For example, neurons in the mammal primary visual cortex that are next to one another are assumed to respond to the same stimulus dimension, such as the tilt of a bar of light. But this is often not the case. Shih-Cheng Yen, Jonathan Baker, and Charles M. Gray, "Heterogeneity in the Responses of Adjacent Neurons to Natural Stimuli in Cat Striate Cortex," *Journal of Neurophysiology* 97, no. 2 (2007): 1326–1341. Neurons like these, which don't respond in the way we expect, have often been ignored.

16. For example, a study may ignore all spikes produced by neurons that fire at a rate more than two standard deviations below that of the neuron with the maximum firing rate.

17. We haven't mentioned noise due to the instruments we are using to study brain activity, such as electrodes, digital cameras, or radio transceivers. In some scientific disciplines such as chemistry, measurement error is the dominant source of noise in an experiment, so estimating its effect is standard practice. But in the brain, other forms of noise like those described above are so big that it is typically not even worth calculating the effect of instrumental noise on data.

18. Bruno A. Olshausen and David J. Field, "What Is the Other 85 Percent of VI Doing?" in *23 Problems in Systems Neuroscience*, ed. J. Leo Van Hemmen and Terrence J. Sejnowski (New York: Oxford University Press, 2005), 182–211.

19. What made the artificial spike distinctive was primarily its high voltage: 10 volts in comparison to the thousandths of a volt of a natural spike. Nevertheless, the unnaturally high voltage wouldn't make the loop any faster. Farran Briggs and W. Martin Usrey, "A Fast, Reciprocal Pathway Between the Lateral Geniculate Nucleus and Visual Cortex in the Macaque Monkey," *Journal of Neuroscience* 27, no. 20 (2007): 5432.

20. Philip Lieberman, *The Unpredictable Species: What Makes Humans Unique* (New York: Princeton University Press, 2013), 2.

21. Ashby identifies computation as the other half of the equation, and he argues that adaptability and rigid computation are not irreconcilable. But while computers do need to be flexible and adapt to current conditions, even fast modern computers are not very good at this. In contrast to Ashby, the internet metaphor views flexibility as a major goal of the protocol of the brain. Flexible communication is not a subfunction of computation. It operates in a different but complementary way. And it supports our unique array of highly flexible cognitive abilities. W. Ross Ashby, *Design for a Brain* (London: Chapman and Hall, 1952), 57.

22. Jiefeng Jiang et al., "An Insula-Frontostriatal Network Mediates Flexible Cognitive Control by Adaptively Predicting Changing Control Demands," *Nature Communications* 6 (2015): 8165.

23. Fergus I. M. Craik and Ellen Bialystok, "Cognition Through the Lifespan: Mechanisms of Change," *Trends in Cognitive Sciences* 10, no. 3 (2006): 131–138.

24. Chess AIs don't use this approach: instead they enumerate all likely states and then compute the statistically optimal one. Our brains are incapable of this, so we use the flexible approach. Moreover, if winning were the only reason humans played chess, we would have given up the game decades ago when computers overtook human competitors. It is rather the exercise of one's cognitive control—and the mental simulation of our opponent's cognitive control process—that make chess enjoyable.

25. Alexander Schlegel, Prescott Alexander, and Peter U. Tse, "Information Processing," 295. This idea originates with Bernard Baars, who termed it the *global workspace*: Bernard J. Baars, "In the Theatre of Consciousness: Global Workspace Theory, a Rigorous Scientific Theory of Consciousness," *Journal of Consciousness Studies* 4, no. 4 (1997): 292–309.

26. Quoted in J. Hadamard, *The Psychology of Invention in the Mathematical Field* (New York: Dover, 1954), 142; Alexander Schlegel, Prescott Alexander, and Peter U. Tse, "Information Processing in the Mental Workspace Is Fundamentally Distributed," *Journal of Cognitive Neuroscience* 28, no. 2 (2016): 295.

27. Fiery Cushman, "Rationalization Is Rational," *Behavioral and Brain Sciences* 43, e28 (2019): 1.

28. Search is just an overlay or an exploitation of latent information related to communication infrastructure in addressing systems. Past proposals in this vein, though intriguing, have ignored the communication infrastructure underlying search. Thomas L. Griffiths, Mark Steyvers, and Alana Firl, "Google and the Mind: Predicting Fluency with PageRank," *Psychological Science* 18, no. 12 (2007): 1069–1076. See also Francisco Varela, Jean-Philippe Lachaux, Eugenio Rodriguez, and Jacques Martinerie, "The Brainweb: Phase Synchronization and Large-Scale Integration," *Nature Reviews Neuroscience* 2, no. 4 (2001): 229–239.

29. Marco Altamirano and Brian N. Mathur, "Where Is My Mind?" *Nautilus* 79, December 19, 2019, http://nautil.us/issue/79/catalysts/where-is-my-mind.

30. In the study of motor systems, the aim has traditionally been to understand how a particular stimulus leads to a particular set of motor responses in the nervous system. In other words, the idea is to understand motor computations, as manifest in brain activity. But it is increasingly understood that much of the activity in motor areas of the brain is not specifically related to the next action to be performed. Vishwanathan Mohan, Ajaz Bhat, and Pietro Morasso, "Muscleless Motor Synergies and Actions Without Movements: From Motor Neuroscience to Cognitive Robotics," *Physics of Life Reviews* (2018).

31. Thomas Weiss et al., "Rapid Functional Plasticity of the Somatosensory Cortex After Finger Amputation," *Experimental Brain Research* 134, no. 2 (2000): 199–203. A more recent experiment in humans puts further emphasis on the fact that signals are re-routed on brain networks almost immediately, with little if any change in connectivity. Neuroscientists put healthy volunteers' arms in a cast for two weeks and measured functional and structural connectivity before, during, and after casting. Disused parts of the brain associated with the casted arm's movements and sensations got a lot less traffic right after casting, then went back to normal just after the cast was removed, and did so over essentially the same network of connections. Dillan J. Newbold, Timothy O. Laumann, Catherine R. Hoyt, Jacqueline M. Hampton, David F. Montez, Ryan V. Raut, Mario Ortega et al., "Plasticity and Spontaneous Activity Pulses in Disused Human Brain Circuits," *Neuron* 107, no. 3 (2020): 580–589.

32. Lothar Spillmann, Birgitta Dresp-Langley, and Chia-huei Tseng, "Beyond the Classical Receptive Field: The Effect of Contextual Stimuli," *Journal of Vision* 15, no. 9 (2015): article 7, 1–23.

33. Alex Fornito, Andrew Zalesky, and Edward Bullmore, *Fundamentals of Brain Network Analysis* (New York: Academic Press, 2016), 251.

34. Laurie Von Melchner, Sarah L. Pallas, and Mriganka Sur, "Visual Behaviour Mediated by Retinal Projections Directed to the Auditory Pathway," *Nature* 404, no. 6780 (2000): 871–876.

35. Gerald Edelman's neural Darwinism framework, for example, is a fairly mainstream account of brain function that rejects the information-processing imperative of the computer metaphor. In this view, cells instead compete among themselves to achieve the best behavioral outcome. Gerald M. Edelman, "Neural Darwinism: Selection and Reentrant Signaling in Higher Brain Function," *Neuron* 10, no. 2 (1993): 115–125. György Buzsáki's grand theory about the brain, detailed in his excellent 2019 book *The Brain from Inside Out*, likewise rejects the computer metaphor in favor of a more network-centered view. György Buzsáki, *The Brain from Inside Out* (New York: Oxford University Press, 2019), 12–13.

36. A review paper in this area points out that "out of the myriad of possible stimuli that can activate a neuron, only a few are likely to be present at any given time." Tim P. Vogels and L. F. Abbott. "Gating Multiple Signals Through Detailed Balance of Excitation and Inhibition in Spiking Networks." *Nature Neuroscience* 12, no. 4 (2009): 488.

37. Alwyn Scott, *Neurophysics* (New York: John Wiley, 1977), 183. See also Alwyn Scott, *Stairway to the Mind: The Controversial New Science of Consciousness* (Berlin: Springer, 1995).

38. Stephen G. Waxman, "Regional Differentiation of the Axon: A Review with Special Reference to the Concept of the Multiplex Neuron," *Brain Research* 47, no. 2 (1972): 269–288.

39. Sheheitli and Jirsa, "Mathematical Model."

40. Scott, *Neurophysics*, 183. Another possible scheme of this kind involves neuromodulatory signals from other neurons. Christof Koch and Tomaso Poggio, "Biophysics of Computation: neurons, synapses and membranes," *Synaptic Function*. Eds. Edelman GM, Gall WE, Cowan WM (1987), 637–697.

41. Citing Scott, a recent review argued that "understanding these dynamics remains of primary interest in the field of neurophysics." Evelyn

Tang and Danielle S. Bassett, "Colloquium: Control of Dynamics in Brain Networks," *Reviews of Modern Physics* 90, no. 3 (2018): 031003–2.

42. Scott, *Neurophysics*, 183.

43. Anthony J. Bell, "Towards a Cross-Level Theory of Neural Learning," *AIP Conference Proceedings* 954 (2007): 56–73.

44. Caroline Möller et al., "Glial Cells for Information Routing?" *Cognitive Systems Research* 8, no. 1 (2007): 28–35.

45. Von der Malsburg is well placed to make this proposal: the rule of thumb that any neuron is about three hops from any other—which predates the connectomics era—is often attributed to him; see: Wiskott and von der Malsburg, "Face Recognition," 1.

46. Philippe Morquette et al., "An Astrocyte-Dependent Mechanism for Neuronal Rhythmogenesis," *Nature Neuroscience* 18, no. 6 (2015): 844.

47. Seiji Matsuda et al., "Phylogenetic Investigation of Dogiel's Pericellular Nests and Cajal's Initial Glomeruli in the Dorsal Root Ganglion," *Journal of Comparative Neurology* 491, no. 3 (2005): 235.

48. P. Somogyi et al., "Synaptic Connections of Morphologically Identified and Physiologically Characterized Large Basket Cells in the Striate Cortex of Cat," *Neuroscience* 10, no. 2 (1983): 261.

49. See, for example, Mircea Steriade and Denis Paré, *Gating in Cerebral Networks* (Cambridge: Cambridge University Press, 2007).

50. Tim Gollisch and Markus Meister, "Eye Smarter Than Scientists Believed: Neural Computations in Circuits of the Retina," *Neuron* 65, no. 2 (2010): 150–164.

51. Thomas Gisiger and Mounir Boukadoum, "Mechanisms Gating the Flow of Information in Cortex: What They Might Look Like and What Their Uses May Be," *Frontiers in Computational Neuroscience* 5 (2011): 1.

52. Buzsáki, *The Brain from Inside Out*, 10–12.

53. Hamid R. Noori et al., "A Multiscale Cerebral Neurochemical Connectome of the Rat Brain," *PLoS Biology* 15, no. 7 (2017): e2002612.

54. The locus coeruleus sends information to "virtually every part of the CNS [central nervous system]." Walter J. Hendelman, *Atlas of Functional Neuroanatomy* (Boca Raton, FL: Taylor & Francis Group, 2016), 70.

55. Noori et al., "A Multiscale Cerebral Neurochemical Connectome."

56. Steriade and Paré, *Gating in Cerebral Networks*, 2, 5.

57. Francis C. Crick and Christof Koch, "What Is the Function of the Claustrum?" *Philosophical Transactions of the Royal Society B: Biological Sciences* 360, no. 1458 (2005): 1271–1279.

58. Sara Reardon, "A Giant Neuron Found Wrapped Around Entire Mouse Brain," *Nature* 543, no. 7643 (March 2017): 14–15.

59. Michael W. Cole, Jeremy R. Reynolds, Jonathan D. Power, Grega Repovs, Alan Anticevic, and Todd S. Braver, "Multi-task Connectivity Reveals Flexible Hubs for Adaptive Task Control," *Nature Neuroscience* 16, no. 9 (2013): 1348–1355.

60. Raphael T. Gerraty et al., "Dynamic Flexibility in Striatal-Cortical Circuits Supports Reinforcement Learning," *Journal of Neuroscience* 38, no. 10 (2018): 2442–2453.

61. Barbara J. Hunnicutt et al., "A Comprehensive Excitatory Input Map of the Striatum Reveals Novel Functional Organization," *Elife* 5 (2016): e19103.

62. This capacity has been termed neural re-use. Michael L. Anderson, "Neural Reuse: A Fundamental Organizational Principle of the Brain," *Behavioral and Brain Sciences* 33, no. 4 (2010): 245.

63. Iva Reichova and S. Murray Sherman. "Somatosensory Corticothalamic Projections: Distinguishing Drivers from Modulators," *Journal of Neurophysiology* 92, no. 4 (2004): 2185–2197.

64. Charles M. Gray and Wolf Singer, "Stimulus-Specific Neuronal Oscillations In Orientation Columns of Cat Visual Cortex," *Proceedings of the National Academy of Sciences* 86, no. 5 (1989): 1698–1702.

65. Pascal Fries, "A Mechanism for Cognitive Dynamics: Neuronal Communication Through Neuronal Coherence," *Trends in Cognitive Sciences* 9, no. 10 (2005): 474. Similar arguments, including references to the analogy of the worldwide web, are elucidated in Varela et al., "Brainweb."

66. The offset in time between a rhythm in one area and the same rhythm in another area means that the two areas are not active at exactly the same time. Since neurons are relatively slow at passing signals, this offset is inevitable.

67. Michel Kerszberg, "Genes, Neurons and Codes: Remarks on Biological Communication," *Bioessays* 25, no. 7 (2003): 699–708.

68 Agostina Palmigiano et al., "Flexible Information Routing by Transient Synchrony," *Nature Neuroscience* 20, no. 7 (2017): 1014; Marije ter Wal and Paul H. Tiesinga, "Phase Difference Between Model Cortical

Areas Determines Level of Information Transfer," *Frontiers in Computational Neuroscience* 11 (2017): 6.

69. The frequency of oscillation depends on the number of neurons being synchronized within a given brain. In small chunks of the brain (within one cortical area), activity oscillates at a very high rate (30–100 pulses per second). When neurons from different cortical areas synchronize, they do so at slower rates (15–25 pulses per second). György Buzsáki, *Rhythms of the Brain* (New York: Oxford University Press, 2006), 151–152.

70. The choice of protocol involves trade-offs. Since it was designed for sparse, asynchronous communication, the internet remains less than ideal for dense, synchronous signals such as real-time video chat. But its fundamental flexibility makes it at least "good enough."

71. Monotremes like the echidna and marsupials like the platypus lack a corpus callosum. Jon H. Kaas, "Reconstructing the Organization of Neocortex of the First Mammals and Subsequent Modifications," in *Evolution of Nervous Systems*, ed. Jon H. Kaas and Leah A. Krubitzer (Amsterdam: Academic Press, 2007), 3: 28.

72. In the shrew, there is 10 percent callosal myelination, while in the macaque the figure is 70 percent. Samuel S-H. Wang, "Functional Tradeoffs in Axonal Scaling: Implications for Brain Function," *Brain, Behavior and Evolution* 72, no. 2 (2008): 163.

73. See also related discussions of the scaling of computer architecture in relation to brains: Valeriu Beiu and Walid Ibrahim, "Does the Brain Really Outperform Rent's Rule?" in *2008 IEEE International Symposium on Circuits and Systems* (New York: The Printing House, 2008), 640–643.

74. Arun S. Mahadevan et al., "Living Neural Networks: Dynamic Network Analysis of Developing Neural Progenitor Cells," *bioRxiv* (2017): 055533.

7. CRITIQUE OF THE INTERNET METAPHOR

1. Eve Marder, "Opening Lecture," Bernstein Conference on Computational Neuroscience, September 18, 2019, Berlin.

2. Santiago Ramón y Cajal, *Advice for a Young Investigator*, trans. Neely Swanson and Larry Swanson, (Cambridge, Ma.: MIT Press, 2004), 3.

3. Eric Newman et al., eds., *The Beautiful Brain: The Drawings of Santiago Ramón y Cajal* (New York: Abrams, 2017).

4. Zoltán Nádasdy et al., "Replay and Time Compression of Recurring Spike Sequences in the Hippocampus," *Journal of Neuroscience* 19, no. 21 (1999): 9497–9507.

5. Dion Khodagholy, Jennifer N. Gelinas, and György Buzsáki, "Learning-Enhanced Coupling Between Ripple Oscillations in Association Cortices and Hippocampus," *Science* 358, no. 6361 (2017): 369–372.

6. This is called the cocktail party problem.

7. More recent research by Buzsáki and colleagues has suggested that neurons in this part of the brain show a substantial range in their degree of flexibility. At one end of the spectrum are "plastic" cells that can vary their pattern of signal passing, and at the other end are "rigid" cells that largely stick to a single pattern. Andres D. Grosmark and György Buzsáki, "Diversity in Neural Firing Dynamics Supports Both Rigid and Learned Hippocampal Sequences," *Science* 351, no. 6280 (2016): 1440–1443.

8. Brains are capable of fine-tuning their connectivity to send signals over routes that take a very precise amount of time. See, for example, Giorgio M. Innocenti, Marie Carlén, and Tim Dyrby, "The Diameters of Cortical Axons and Their Relevance to Neural Computing in Axons and Brain Architecture," in *Axons and Brain Architecture*, ed. Kathleen Rockland (New York: Academic Press, 2016), 317–355.

9. Barbara L. Finlay and Richard B. Darlington, "Linked Regularities in the Development and Evolution of Mammalian Brains," *Science* 268, no. 5217 (1995): 1578–1584.

10. Fang Fang, Daniel Kersten, and Scott Murray, "Perceptual Grouping and Inverse fMRI Activity Patterns in Human Visual Cortex," *Journal of Vision* 8, no. 7 (2008): 2.

11. John T. Arsenault et al., "Attention Shifts Recruit the Monkey Default Mode Network," *Journal of Neuroscience* 38, no. 5 (2018): 1202–1217.

12. Connectivity can be measured with tracers in human brains extracted after death, but major challenges prevent large-scale studies of this kind. Tracers work best when injected while the brain's owner is still alive—obviously off-limits in humans—and ideally tracing should happen soon after death. Since donated brains are rare to begin with, these factors make chemical tracing almost impossible in humans.

13. Klaus H. Maier-Hein et al., "The Challenge of Mapping the Human Connectome Based on Diffusion Tractography," *Nature Communications* 8, no. 1 (2017): 1349.

14. Structural imaging approaches detect bundles of axons in millimeter-sized cubes of brain tissue. They can measure the most likely orientation (in 3D space) at which axons enter the cube and the most likely orientation at which axons leave the cube. The problem comes if more than one bundle of axons crosses the cube and enters and exits in different directions. In this case, which is common in the brain, many axons will be ignored in most standard structural imaging approaches.

15. Penelope Kale, Andrew Zalesky, and Leonardo L. Gollo, "Estimating the Impact of Structural Directionality: How Reliable Are Undirected Connectomes?" *Network Neuroscience* 2, no. 2 (2018): 259–284.

16. Peter Sheridan Dodds, Duncan J. Watts, and Charles F. Sabel, "Information Exchange and the Robustness of Organizational Networks," *Proceedings of the National Academy of Sciences* 100, no. 21 (2003): 12516–12521.

17. Note that the term *sparse* is also used to describe the connectivity of a network. Sparse connectivity means that each node is connected to only a small fraction of the other nodes. Here the term *sparse* refers to dynamic activity, which means it resembles character usage in written Chinese (see chapter 3).

18. Takashi Yoshida and Kenichi Ohki, "Natural Images Are Reliably Represented by Sparse and Variable Populations of Neurons in Visual Cortex," *Nature Communications* 11, no. 1 (2020): 1–19.

19. Some have hypothesized that buffers exist in neural circuits of the prefrontal cortex, an area involved in working memory. Working memory has an obvious need to maintain storage over short time scales. See Patricia S. Goldman-Rakic, "Regional and Cellular Fractionation of Working Memory," *Proceedings of the National Academy of Sciences* 93, no. 24 (1996): 13473–13480; and Shintaro Funahashi, "Functions of Delay-Period Activity in the Prefrontal Cortex and Mnemonic Scotomas Revisited," *Frontiers in Systems Neuroscience* 9 (2015): 2.

20. Emma Strubell, Ananya Ganesh, and Andrew McCallum, "Energy and Policy Considerations for Deep Learning in NLP," *arXiv: 1906.02243* (2019).

21. Vivienne Ming, "Human Insight Remains Essential to Beat the Bias of Algorithms," *Financial Times*, December 3, 2019, https://www.ft.com/content/59520726-docs-11e9-b018-ca4456540ea6.

22. Jonathan C. Horton and Daniel L. Adams, "The Cortical Column: A Structure Without a Function," *Philosophical Transactions of the Royal Society B: Biological Sciences* 360, no. 1456 (2005): 837–862.

23. David Quammen, *The Tangled Tree* (New York: Simon and Schuster, 2018), 297–304.

8. THE INTERNET METAPHOR IN ACTION: EMERGING MODELS AND NEW TECHNOLOGIES

1. Nikola T. Markov et al., "A Weighted and Directed Interareal Connectivity Matrix for Macaque Cerebral Cortex," *Cerebral Cortex* 24, no. 1 (2014): 17–36; Seung Wook Oh et al., "A Mesoscale Connectome of the Mouse Brain," *Nature* 508, no. 7495 (2014): 207.

2. C. Shan Xu et al., "A Connectome of the Adult *Drosophila* Central Brain," *bioRxiv*, 2020, https://www.biorxiv.org/content/biorxiv/early/2020/01/21/2020.01.21.911859.full.pdf.

3. In mammal brains, slices are made using a microtome, which resembles a meat cutter at the deli counter but is capable of making extremely thin slices. They are then imaged with a microscope. In the fly, the brain is small enough that electron microscopes can be used. A whole slice can be imaged and then vaporized using an ion beam. By repeatedly imaging and then zapping away layers of the tiny brain, one can construct a full 3D set of neural connections. Unfortunately, mammal brains are too big for this approach to be workable at present.

4. Moritz Helmstaedter et al., "Connectomic Reconstruction of the Inner Plexiform Layer in the Mouse Retina," *Nature* 500, no. 7461 (2013): 168.

5. Recent estimates suggest that deducing when a given brain disorder can be associated with altered structure using MR imaging requires more than 10,000 participants: Scott Marek, Brenden Tervo-Clemmens, Finnegan J. Calabro, David F. Montez, Benjamin P. Kay, Alexander S. Hatoum, Meghan Rose Donohue et al., "Towards Reproducible Brain-Wide Association Studies." *bioRxiv* (2020). See also Michael N. Hallquist and Frank G. Hillary, "Graph Theory Approaches to Functional Network Organization in Brain Disorders: A Critique for a Brave New Small-World," *Network Neuroscience* 3, no. 1 (2018): 1–26.

6. Bertha Vázquez-Rodríguez et al., "Gradients of Structure–Function Tethering Across Neocortex," *Proceedings of the National Academy of Sciences* 116, no. 42 (2019): 21219–21227.

7. Caio Seguin, Ye Tian, and Andrew Zalesky, "Network Communication Models Improve the Behavioral and Functional Predictive Utility of the Human Structural Connectome," Network Neuroscience 4 no. 4 (2020): 980–1006.

8. They literally wrote the book on machine-to-machine communication. Vojislav B. Mišić and Jelena Mišić, eds., *Machine-to-Machine Communications: Architectures, Technology, Standards, and Applications* (Boca Raton, FL: CRC Press, 2014).

9. The existence of connections was based on firm evidence using neuronal tracer molecules (no imaging data are included in the collated data). This was a monumental achievement of synthesis, not least because the thousands of studies invariably define brain regions in different ways. Klaas Enno Stephan, "The History of CoCoMac," *Neuroimage* 80 (2013): 46–52.

10. Bratislav Mišić, Olaf Sporns, and Anthony R. McIntosh, "Communication Efficiency and Congestion of Signal Traffic in Large-Scale Brain Networks," *PLoS Computational Biology* 10, no. 1 (2014).

11. R. Eckert, Y. Naitoh, and K. Friedman, "Sensory Mechanisms in Paramecium. I," *Journal of Experimental Biology* 56 (1972): 683–694.

12. Sharla P. Boehm and Paul Baran, "On Distributed Communications: II. Digital Simulation of Hot-Potato Routing in a Broadband Distributed Communications Network," *Memorandum of the RAND Corporation Prepared for United States Air Force* (1964). This solution also requires that there are no dead ends on the network: all nodes must be both senders and receivers. This condition is met on the internet and probably also on most parts of the brain's connectome. Hamid R. Noori et al., "A Multiscale Cerebral Neurochemical Connectome of the Rat Brain," *PLoS Biology* 15, no. 7 (2017): e2002612.

13. Andrea Avena-Koenigsberger et al., "A Spectrum of Routing Strategies for Brain Networks," *PLoS Computational Biology* 15, no. 3 (2019): e1006833. This paper was among the first to bring the term *routing* into the mainstream of research on brain networks.

14. Caio Seguin, Martijn P. Van Den Heuvel, and Andrew Zalesky, "Navigation of Brain Networks," *Proceedings of the National Academy of Sciences* 115, no. 24 (2018): 6297–6302.

15. Yan Hao and Daniel Graham, "Creative Destruction: Sparse Activity Emerges on the Mammal Connectome Under a Simulated Routing Protocol with Collisions and Redundancy," *Network Neuroscience* 4, no. 4 (2020), 1055–1071. In reality, destruction of all colliding messages is probably too strong an assumption. Simultaneous input at multiple dendrites is probably necessary to make the cell fire, though a small subset of inputs is probably sufficient. Nevertheless, colliding signals can and do regularly destroy each other in the brain; see, for example, Shira Sardi et al., "New Types of Experiments Reveal That a Neuron Functions as Multiple Independent Threshold Units," *Scientific Reports* 7, no. 1 (2017): 1–17. Destructive collisions are also predicted by mechanistic models of single-neuron interactions (based on the Hodgkin-Huxley equations; Alwyn Scott, *Neurophysics* [New York: John Wiley, 1977], 184); by gating models (Tim Gollisch and Markus Meister, "Eye Smarter Than Scientists Believed: Neural Computations in Circuits of the Retina," *Neuron* 65 [2010]: 153); and by other mechanisms (Albert Gidon et al., "Dendritic Action Potentials and Computation in Human Layer 2/3 Cortical Neurons," *Science* 367, no. 6473 [2020]: 83–87). While most computer metaphor–based models of neural systems see collisions as desirable summations, the perspective from communications systems and the internet is that collisions need to be managed. Hao and I therefore examined how an information spreading model could aid this goal in the context of destructive collisions.

16. Markov et al., "A Weighted and Directed Interareal Connectivity," 17–36; Oh et al., "A Mesoscale Connectome of the Mouse Brain," 207.

17. Dirk Bucher, "Contribution of Axons to Short-Term Dynamics of Neuronal Communication," in *Axons and Brain Architecture*, ed. Kathleen Rockland (London: Academic Press, 2016).

18. Laura E. Suárez et al., "Linking Structure and Function in Macroscale Brain Networks," *Trends in Cognitive Sciences* 24, no. 4 (2020): 302–315.

19. Daniel Graham, Andrea Avena Koenigsberger and Bratislav Mišić, "Focus Feature: Network Communication in the Brain," *Network Neuroscience* 4, no. 4: 976–979.

20. Steve B. Furber et al., "The SpiNNaker Project," *Proceedings of the IEEE* 102, no. 5 (2014): 652–665.

21. See, for example, Ed Bullmore and Olaf Sporns, "Complex Brain Networks: Graph Theoretical Analysis of Structural and Functional Systems." *Nature Reviews Neuroscience* 10, no. 3 (2009): 186–198.

22. Elizabeth N. Davison et al., "Brain Network Adaptability Across Task States," *PLoS Computational Biology* 11, no. 1 (2015).

23. James H. Marshel et al., "Cortical Layer–Specific Critical Dynamics Triggering Perception," *Science* 365, no. 6453 (2019): eaaw5202.

24. There are related biotechnological tools that deploy neurotransmitters in highly specific ways to individual cells: neurotransmitter molecules can be kept inactive in "cages" within the neuron until the experimenter releases them. This has the effect of turning on selected cells. The connections among selected neurons can also be traced after their activity is measured. This is a promising approach. Daniel A. Llano and S. Murray Sherman, "Differences in Intrinsic Properties and Local Network Connectivity of Identified Layer 5 and Layer 6 Adult Mouse Auditory Corticothalamic Neurons Support a Dual Corticothalamic Projection Hypothesis," *Cerebral Cortex* 19, no. 12 (2009): 2810–2826.

25. It is possible that there will be a sharp increase in traffic at very high levels of stimulation.

26. David Kleinfeld et al., "Can One Concurrently Record Electrical Spikes from Every Neuron in a Mammalian Brain?" *Neuron* 103, no. 6 (2019): 1005–1015.

27. Edward M. Callaway and Anupam K. Garg, "Brain Technology: Neurons Recorded en Masse," *Nature* 551, no. 7679 (2017): 172.

28. Raju Tomer et al., "Advanced CLARITY for Rapid and High-Resolution Imaging of Intact Tissues," *Nature Protocols* 9, no. 7 (2014): 1682.

9. THE INTERNET METAPHOR, AI, AND US

1. Arvind Narayanan, "How to Recognize AI Snake Oil," Center for Information Technology Policy, Princeton University, https://www.cs.princeton.edu/~arvindn/talks/MIT-STS-AI-snakeoil.pdf.

2. David Spiegelhalter, "Should We Trust Algorithms?" *Harvard Data Science Review* 2, no. 1 (2020).

3. Amir Rosenfeld, Richard Zemel, and John K. Tsotsos, "The Elephant in the Room," *arXiv: 1808.03305* (2018).

4. H. Sebastian Seung and Raphael Yuste, "Neural Networks," in *Principles of Neural Science*, ed. Eric R. Kandel and Thomas J. Jessel (New York: McGraw-Hill, 2013), 1581.

5. Melanie Boly et al., "Consciousness in Humans and Non-human Animals: Recent Advances and Future Directions," *Frontiers in Psychology* 4 (2013): 625.

6. This fact is a key problem for the computer metaphor, especially in relation to perception. As Brette writes, "Current evidence indicates that the activity of neurons is sensitive to stimulus properties . . . but cannot be considered as context-free symbols that stand for the corresponding properties. . . . Can neural coding theories of perception accommodate for this fact? It would require that in every context, changes in encoding (stimulus-response properties) are exactly mirrored by changes in decoding (computations performed on neural activity, e.g. the "simple neural readout"). No mechanism has been proposed to achieve this." Romain Brette, "Is Coding a Relevant Metaphor for the Brain?" *Behavioral and Brain Sciences* 42 (2019): 1–44.

7. Oliver Sacks, "In the River of Consciousness," *New York Review of Books* 51, no. 1 (2004): 41–45.

8. Valtteri Arstila and Dan Lloyd, *Subjective Time: The Philosophy, Psychology, and Neuroscience of Temporality* (Cambridge, MA: MIT Press, 2014), 159.

9. Joseph Hindman et al., "Thalamic Strokes That Severely Impair Arousal Extend into the Brainstem," *Annals of Neurology* 84, no. 6 (2018): 926–930.

10. Sophia Wood, "How Mexican App Bridgefy Is Connecting Protesters in Hong Kong," *LatAm List*, August 23, 2019, https://latamlist.com/2019/08/23/how-mexican-app-bridgefy-is-connecting-protesters-in-hong-kong/.

11. Vojislav B. Mišić and Jelena Mišić, eds., *Machine-to-Machine Communications: Architectures, Technology, Standards, and Applications* (Boca Raton, FL: CRC Press, 2014).

12. The technology firm Neuralink, backed by Musk, is promising to develop electrodes that Musk says can "read from and write to" the brain. Although the company "is under no illusions" that this is an easy task, there is every indication that its approach will be guided exclusively by the computer metaphor. Patrick McGee, "Elon Musk–backed Neuralink

Unveils Brain-Implant Technology," *Financial Times*, July 17, 2019, https://www.ft.com/content/144ba3b4-a85a-11e9-984c-fac8325aaa04.

13. For a lively and sobering account of the transhumanism movement, see Mark O'Connell, *To Be a Machine: Adventures Among Cyborgs, Utopians, Hackers, and the Futurists Solving the Modest Problem of Death* (New York: Anchor, 2018).

14. Joel Zylberberg and Michael Robert DeWeese, "Sparse Coding Models Can Exhibit Decreasing Sparseness While Learning Sparse Codes for Natural Images," *PLoS Computational Biology* 9, no. 8 (2013): e1003182.

AFTERWORD

1. A recent review of the evolution of the mammal cortex used the term *great flexibility* twice and the term *extremely flexible* once in the span of three pages when describing the mammal cortex. Jon H. Kaas, "Reconstructing the Organization of Neocortex of the First Mammals and Subsequent Modifications," in *Evolution of Nervous Systems*, ed. Jon H. Kaas and Leah A. Krubitzer (Amsterdam: Academic Press, 2007), 3: 30–32.

2. Tethered by a wire to a tiny stage, the flies were taught to head toward a particular shape—the letter T. If they went toward other shapes, the flies were punished by being zapped with an infrared laser. Amazingly, the flies showed invariance to the position of the T: having been trained with the T falling on one part of the retina, they zoomed toward a T falling on an entirely different part of the retina without training. As we saw in chapter 1, invariance of this kind is precisely the kind of problem that internet-like routing is good at solving. Shiming Tang et al., "Visual Pattern Recognition in *Drosophila* Is Invariant for Retinal Position," *Science* 305, no. 5686 (2004): 1020–1022.

3. C. Shan Xu et al., "A Connectome of the Adult *Drosophila* Central Brain," *bioRxiv* (2020). https://www.biorxiv.org/content/10.1101/2020.01.21.911859v1.full.pdf.

BIBLIOGRAPHY

Abbott, Larry F., and Frances S. Chance. "Rethinking the Taxonomy of Visual Neurons." *Nature Neuroscience* 5, no. 5 (2002): 391.

Alivisatos, A. Paul, Miyoung Chun, George M. Church, Ralph J. Greenspan, Michael L. Roukes, and Rafael Yuste. "The Brain Activity Map Project and the Challenge of Functional Connectomics." *Neuron* 74, no. 6 (2012): 970–974.

Altamirano, Marco, and Brian N. Mathur. "Where Is My Mind?" *Nautilus* 79, December 19, 2019. http://nautil.us/issue/79/catalysts/where-is-my-mind.

Anderson, Michael L. "Neural Reuse: A Fundamental Organizational Principle of the Brain." *Behavioral and Brain Sciences* 33, no. 4 (2010): 245.

Anderson, Philip W. "More Is Different." *Science* 177, no. 4047 (1972): 393–396.

Arsenault, John T., Natalie Caspari, Rik Vandenberghe, and Wim Vanduffel. "Attention Shifts Recruit the Monkey Default Mode Network." *Journal of Neuroscience* 38, no. 5 (2018): 1202–1217.

Arstila, Valtteri, and Dan Lloyd. *Subjective Time: The Philosophy, Psychology, and Neuroscience of Temporality*. Cambridge, MA: MIT Press, 2014.

Ashby, W. Ross. *Design for a Brain*. London: Chapman and Hall, 1952.

Attwell, David, and S. B. Laughlin. "An Energy Budget for Signaling in the Grey Matter of the Brain." *Journal of Cerebral Blood Flow and Metabolism* 21, no. 10 (2001): 1133–1145.

Avena-Koenigsberger, Andrea, Xiaoran Yan, Artemy Kolchinsky, Martijn Van Den Heuvel, Patric Hagmann, and Olaf Sporns. "A Spectrum of Routing Strategies for Brain Networks." *PLoS Computational Biology* 15, no. 3 (2019): E1006833.

Azevedo, Frederico A. C., Ludmila R. B. Carvalho, Lea T. Grinberg, José Marcelo Farfel, Renata EL Ferretti, Renata E. P. Leite, Wilson Jacob Filho, Roberto Lent, and Suzana Herculano-Houzel. "Equal Numbers of Neuronal and Nonneuronal Cells Make the Human Brain an Isometrically Scaled-Up Primate Brain." *Journal of Comparative Neurology* 513, no. 5 (2009): 532–541.

Baars, Bernard J. "In the Theatre of Consciousness: Global Workspace Theory, a Rigorous Scientific Theory of Consciousness." *Journal of Consciousness Studies* 4, no. 4 (1997): 292–309.

Backstrom, Lars, Paolo Boldi, Marco Rosa, Johan Ugander, and Sebastiano Vigna. "Four Degrees of Separation." In *Proceedings of the Fourth Annual ACM Web Science Conference*, 33–42. New York: Association for Computing Machinery, 2012.

Baddeley, Roland, Larry F. Abbott, Michael C. A. Booth, Frank Sengpiel, Tobe Freeman, Edward A. Wakeman, and Edmund T. Rolls. "Responses of Neurons in Primary and Inferior Temporal Visual Cortices to Natural Scenes." *Proceedings of the Royal Society of London B: Biological Sciences* 264, no. 1389 (1997): 1775–1783.

Barlow, Horace Basil. "Possible principles underlying the transformation of sensory messages" In *Sensory Communication* ed W. A. Rosenblith." Cambridge, MA: MIT Press, 1961.

Barlow, Horace Basil. "Why Have Multiple Cortical Areas?" *Vision Research* 26, no. 1 (1986): 81–90.

Bassett, Danielle S., and Edward T. Bullmore. "Small-World Brain Networks Revisited," *Neuroscientist* 23, no. 5 (2017): 499–516.

Bassett, Danielle S. and Michael S. Gazzaniga. "Understanding Complexity in the Human Brain." *Trends in Cognitive Sciences* 15, no. 5 (2011): 200–209.

Beiu, Valeriu, and Walid Ibrahim. "Does the Brain Really Outperform Rent's Rule?" In *2008 IEEE International Symposium on Circuits and Systems*, 640–643. New York: The Printing House, 2008.

Bell, Anthony J. "Levels and Loops: The Future of Artificial Intelligence and Neuroscience." *Philosophical Transactions of the Royal Society of London B: Biological Sciences* 354, no. 1392 (1999): 2013–2020.

Bell, Anthony J., and Terrence J. Sejnowski. "The 'Independent Components' of Natural Scenes Are Edge Filters." *Vision Research* 37, no. 23 (1997): 3327–3338.

Binzegger, Tom, Rodney J. Douglas, and Kevan A. C. Martin. "Topology and Dynamics of the Canonical Circuit of Cat V1." *Neural Networks* 22, no. 8 (2009): 1071–1078.

Boehm, Sharla P. and Paul Baran. "On Distributed Communications: II. Digital Simulation of Hot-Potato Routing in a Broadband Distributed Communications Network." *Memorandum of the RAND corporation prepared for United States Air Force* (1964).

Boly, Melanie, Anil K. Seth, Melanie Wilke, Paul Ingmundson, Bernard Baars, Steven Laureys, David Edelman, and Naotsugu Tsuchiya. "Consciousness in Humans and Non-Human Animals: Recent Advances and Future Directions." *Frontiers in Psychology* 4 (2013): 625.

Brette, Romain. "Is Coding a Relevant Metaphor for the Brain?" *Behavioral and Brain Sciences* 42 (2019): 1–44.

Briggs, Farran, and W. Martin Usrey. "A Fast, Reciprocal Pathway Between the Lateral Geniculate Nucleus and Visual Cortex in the Macaque Monkey." *Journal of Neuroscience* 27, no. 20 (2007): 5431–5436.

Bucher, Dirk. "Contribution of Axons to Short-Term Dynamics of Neuronal Communication." In *Axons and Brain Architecture*, ed. Kathleen Rockland. London: Academic Press, 2016.

Bullmore, Ed, and Olaf Sporns. "Complex Brain Networks: Graph Theoretical Analysis of Structural and Functional Systems." *Nature Reviews Neuroscience* 10, no. 3 (2009): 186–198.

Buonomano, Dean. *Brain Bugs*. New York: Norton, 2011.

Burton, Geoffrey J., and Ian R. Moorhead. "Color and Spatial Structure in Natural Scenes." *Applied Optics* 26, no. 1 (1987): 157–170.

Buzsáki, György. *The Brain from Inside Out*. New York: Oxford University Press, 2019.

Buzsáki, György. *Rhythms of the Brain*. New York: Oxford University Press, 2006.

Callaway, Edward M., and Anupam K. Garg. "Brain Technology: Neurons Recorded en Masse." *Nature* 551, no. 7679 (2017): 172.

Chang, Le, and Doris Y. Tsao. "The Code for Facial Identity in the Primate Brain." *Cell* 169, no. 6 (2017): 1013–1028.

Changizi, Mark A., Qiong Zhang, and Shinsuke Shimojo. "Bare Skin, Blood and the Evolution of Primate Colour Vision." *Biology Letters* 2, no. 2 (2006): 217–221.

Charles, Adam S., Mijung Park, J. Patrick Weller, Gregory D. Horwitz, and Jonathan W. Pillow. "Dethroning the Fano Factor: A Flexible, Model-Based Approach to Partitioning Neural Variability." *Neural Computation* 30, no. 4 (2018): 1012–1045.

Churchland, Patricia S. and Terrence J. Sejnowski. *The Computational Brain.* Cambridge, MA: MIT Press, 2016.

Cole, Michael W., Jeremy R. Reynolds, Jonathan D. Power, Grega Repovs, Alan Anticevic, and Todd S. Braver. "Multi-task Connectivity Reveals Flexible Hubs for Adaptive Task Control." *Nature Neuroscience* 16, no. 9 (2013): 1348–1355.

Cichy, Radoslaw Martin, Aditya Khosla, Dimitrios Pantazis, Antonio Torralba, and Aude Oliva. "Comparison of Deep Neural Networks to Spatio-Temporal Cortical Dynamics of Human Visual Object Recognition Reveals Hierarchical Correspondence." *Scientific Reports* 6, article 27755 (2016).

Cisek, Paul. "Beyond the Computer Metaphor: Behaviour as Interaction." *Journal of Consciousness Studies* 6, nos. 11–12 (1999): 125–142.

Conway, Flo, and Jim Siegelman. *Dark Hero of the Information Age: In Search of Norbert Wiener, the Father of Cybernetics.* New York: Basic Books, 2006.

Craik, Fergus I. M., and Ellen Bialystok, "Cognition Through the Lifespan: Mechanisms of Change." *Trends in Cognitive Sciences* 10, no. 3 (2006): 131–138.

Crick, Francis C., and Christof Koch, "What Is the Function of the Claustrum?" *Philosophical Transactions of the Royal Society B: Biological Sciences* 360, no. 1458 (2005): 1271–1279.

Cushman, Fiery. "Rationalization Is Rational." *Behavioral and Brain Sciences* 43, e28 (2019): 1–59.

Davison, Elizabeth N., Kimberly J. Schlesinger, Danielle S. Bassett, Mary-Ellen Lynall, Michael B. Miller, Scott T. Grafton, and Jean M. Carlson. "Brain Network Adaptability Across Task States." *PLoS Computational Biology* 11, no. 1 (2015).

Dehmer, Matthias, and Abbe Mowshowitz. "A History of Graph Entropy Measures." *Information Sciences* 181, no. 1 (2011): 57–78.

Descartes, René. *Treatise of Man,* trans. P. R. Sloan. In *The History and Philosophy of Science,* ed. Daniel McKaughan and Holly Vandewall. London: Bloomsbury Academic, 2018.

de Vries, Saskia EJ, Jerome A. Lecoq, Michael A. Buice, Peter A. Groblewski, Gabriel K. Ocker, Michael Oliver, David Feng et al. "A Large-Scale

Standardized Physiological Survey Reveals Functional Organization of the Mouse Visual Cortex." *Nature Neuroscience* 23, no. 1 (2020): 138–151.

Dodds, Peter Sheridan, Duncan J. Watts, and Charles F. Sabel. "Information Exchange and the Robustness of Organizational Networks." *Proceedings of the National Academy of Sciences* 100, no. 21 (2003): 12516–12521.

Eccles, John Carew. *The Physiology of Synapses*. New York: Academic Press, 1964.

Ecker, Alexander S., Philipp Berens, R. James Cotton, Manivannan Subramaniyan, George H. Denfield, Cathryn R. Cadwell, Stelios M. Smirnakis, Matthias Bethge, and Andreas S. Tolias. "State Dependence of Noise Correlations in Macaque Primary Visual Cortex." *Neuron* 82, no. 1 (2014): 235–248.

Eckert, R., Y. Naitoh, and K. Friedman. "Sensory Mechanisms in Paramecium. I." *Journal of Experimental Biology* 56 (1972): 684–694.

Edelman, Gerald M. "Neural Darwinism: Selection and Reentrant Signaling in Higher Brain Function." *Neuron* 10, no. 2 (1993): 115–125.

El Gamal, Abbas, and Young-Han Kim. *Network Information Theory*. Cambridge: Cambridge University Press, 2011.

Engineering and Technology History Wiki. "Versailles Fountains." https://ethw.org/versailles_fountains.

Epsztein, Jérôme, Albert K. Lee, Edith Chorev, and Michael Brecht. "Impact of Spikelets on Hippocampal CA1 Pyramidal Cell Activity During Spatial Exploration." *Science* 327, no. 5964 (2010): 474–477.

Ercsey-Ravasz, Mária, Nikola T. Markov, Camille Lamy, David C. Van Essen, Kenneth Knoblauch, Zoltán Toroczkai, and Henry Kennedy. "A Predictive Network Model of Cerebral Cortical Connectivity Based on a Distance Rule." *Neuron* 80, no. 1 (2013): 184–197.

Fabre-Thorpe, Michèle, Arnaud Delorme, Catherine Marlot, and Simon Thorpe. "A Limit to the Speed of Processing in Ultra-Rapid Visual Categorization of Novel Natural Scenes." *Journal of Cognitive Neuroscience* 13, no. 2 (2001): 171–180.

Faisal, A. Aldo, Luc P. J. Selen, and Daniel M. Wolpert. "Noise in the Nervous System." *Nature Reviews Neuroscience* 9, no. 4 (2008): 292.

Fang, Fang, Daniel Kersten, and Scott Murray. "Perceptual Grouping and Inverse fmri Activity Patterns in Human Visual Cortex." *Journal of Vision* 8, no. 7 (2008): 2.

Field, David J. "Relations Between the Statistics of Natural Images and the Response Properties of Cortical Cells." *Josa a* 4, no. 12 (1987): 2379–2394.

Finlay, Barbara L., and Richard B. Darlington. "Linked Regularities in the Development and Evolution of Mammalian Brains." *Science* 268, no. 5217 (1995): 1578–1584.

Finlay, Barbara L., Luiz Carlos De Lima Silveira, and Andreas Reichenbach. "Comparative Aspects of Visual System Development." In *The Primate Visual System: A Comparative Approach*, ed. Jan Kremers, 37–72. Chichester, UK: John Wiley, 2005.

Fornito, Alex, Andrew Zalesky, and Edward Bullmore. *Fundamentals of Brain Network Analysis*. New York: Academic Press, 2016.

Fox, Margalit. *The Riddle of the Labyrinth: The Quest to Crack an Ancient Code*. New York: HarperCollins, 2013.

Fries, Pascal. "A Mechanism for Cognitive Dynamics: Neuronal Communication Through Neuronal Coherence." *Trends in Cognitive Sciences* 9, no. 10 (2005): 474–480.

Friston, Karl. "The Free-Energy Principle: A Rough Guide to the Brain?" *Trends in Cognitive Sciences* 13, no. 7 (2009): 293–301.

Friston, Karl. "The Free-Energy Principle: A Unified Brain Theory?" *Nature Reviews Neuroscience* 11, no. 2 (2010): 127–138.

Fukushima, Kunihiko. "Neocognitron: A Hierarchical Neural Network Capable of Visual Pattern Recognition." *Neural Networks* 1, no. 2 (January 1988): 119–130.

Funahashi, Shintaro. "Functions of Delay-Period Activity in the Prefrontal Cortex and Mnemonic Scotomas Revisited." *Frontiers in Systems Neuroscience* 9 (2015): 2.

Furber, Steve B., Francesco Galluppi, Steve Temple, and Luis A. Plana. "The SpiNNaker Project." *Proceedings of the IEEE* 102, no. 5 (2014): 652–665.

Gămănuţ, Răzvan, Henry Kennedy, Zoltán Toroczkai, Mária Ercsey-Ravasz, David C. Van Essen, Kenneth Knoblauch, and Andreas Burkhalter. "The Mouse Cortical Connectome, Characterized by an Ultra-Dense Cortical Graph, Maintains Specificity by Distinct Connectivity Profiles." *Neuron* 97, no. 3 (2018): 698–715.

Gerraty, Raphael T., Juliet Y. Davidow, Karin Foerde, Adriana Galvan, Danielle S. Bassett, and Daphna Shohamy. "Dynamic Flexibility in Striatal-Cortical Circuits Supports Reinforcement Learning." *Journal of Neuroscience* 38, no. 10 (2018): 2442–2453.

Gerstner, Wulfram, and Richard Naud. "How Good Are Neuron Models?" *Science* 326, no. 5951 (2009): 379–380.

Ghodrati, Masoud, Seyed-Mahdi Khaligh-Razavi, and Sidney R. Lehky. "Towards Building a More Complex View of the Lateral Geniculate Nucleus: Recent Advances in Understanding Its Role." *Progress in Neurobiology* 156 (2017): 214–255.

Gidon, Albert, Timothy Adam Zolnik, Pawel Fidzinski, Felix Bolduan, Athanasia Papoutsi, Panayiota Poirazi, Martin Holtkamp, Imre Vida, and Matthew Evan Larkum. "Dendritic Action Potentials and Computation in Human Layer 2/3 Cortical Neurons." *Science* 367, no. 6473 (2020): 83–87.

Gillies, James, and R. Cailliau. *How the Web Was Born: The Story of the World Wide Web.* New York: Oxford University Press, 2000.

Gisiger, Thomas, and Mounir Boukadoum. "Mechanisms Gating the Flow of Information in Cortex: What They Might Look Like and What Their Uses May Be." *Frontiers in Computational Neuroscience* 5 (2011): 1.

Gold, Joshua I., and Michael N. Shadlen. "The Neural Basis of Decision Making." *Annual Review of Neuroscience* 30 (July 2007): 535–574.

Goldman-Rakic, Patricia S. "Regional and Cellular Fractionation of Working Memory." *Proceedings of the National Academy of Sciences* 93, no. 24 (1996): 13473–13480.

Goldsmith, Andrea, Michelle Effros, Ralf Koetter, Muriel Medard, Asu Ozdaglar, and Lizhong Zheng. "Beyond Shannon: The Quest for Fundamental Performance Limits of Wireless ad Hoc Networks." *IEEE Communications Magazine* 49, no. 5 (2011): 195–205.

Goris, Robbe L. T., J. Anthony Movshon, and Eero P. Simoncelli. "Partitioning Neuronal Variability." *Nature Neuroscience* 17, no. 6 (2014): 858.

Gould, Stephen Jay. *The Structure of Evolutionary Theory.* Cambridge, MA: Harvard University Press, 2002.

Graham, Daniel J. "Routing in the Brain." *Frontiers in Computational Neuroscience* 8 (2014): 44.

Graham, Daniel, Andrea Avena Koenigsberger and Bratislav Mišic,"Focus Feature: Network Communication in the Brain," Network Neuroscience4, no. 4 (2020): 976–979.

Graham, Daniel J., and David J. Field. "Sparse Coding in the Neocortex." In *Evolution of Nervous Systems*, ed. Jon H. Kaas and Leah A. Krubitzer, 181–187. Vol. 3. Amsterdam: Academic Press, 2007.

Graham, Daniel J., and Daniel Rockmore. "The Packet Switching Brain." *Journal of Cognitive Neuroscience* 23, no. 2 (2011): 267–276.

Gray, Charles M., and Wolf Singer. "Stimulus-Specific Neuronal Oscillations In Orientation Columns of Cat Visual Cortex." *Proceedings of the National Academy of Sciences* 86, no. 5 (1989): 1698–1702.

Griffiths, Thomas L., Mark Steyvers, and Alana Firl. "Google and the Mind: Predicting Fluency with PageRank." *Psychological Science* 18, no. 12 (2007): 1069–1076.

Grosmark, Andres D., and György Buzsáki. "Diversity in Neural Firing Dynamics Supports Both Rigid and Learned Hippocampal Sequences." *Science* 351, no. 6280 (2016): 1440–1443.

Gruber, Howard E., and Paul H. Barrett. *Darwin on Man: A Psychological Study of Scientific Creativity*. London: Wildwood House, 1974.

Hadamard, Jacques. *The Psychology of Invention in the Mathematical Field*. New York: Dover, 1954.

Hagmann, Patric. "From Diffusion MRI to Brain Connectomics." Ph.D. dissertation, EPFL Theses, 2005.

Hallquist, Michael N., and Frank G. Hillary. "Graph Theory Approaches to Functional Network Organization in Brain Disorders: A Critique for a Brave New Small-World." *Network Neuroscience* 3, no. 1 (2018): 1–26.

Hancock, John T. *Cell Signalling*. Oxford: Oxford University Press, 2017.

Hao, Yan, and Daniel Graham. "Creative Destruction: Sparse Activity Emerges on the Mammal Connectome Under a Simulated Routing Protocol with Collisions and Redundancy." *Network Neuroscience* 4, no. 4 (2020): 1055–1071.

Harley, Adam. "An Interactive Node-Link Visualization of Convolutional Neural Networks." International Symposium on Visual Computing, 2015. http://scs.ryerson.ca/~aharley/vis/conv.

Hasantash, Maryam, Rosa Lafer-Sousa, Arash Afraz, and Bevil R. Conway. "Paradoxical Impact of Memory on Color Appearance of Faces." *Nature Communications* 10, no. 1 (2019): 1–10.

Häusser, Michael, and Bartlett Mel. "Dendrites: Bug or Feature?" *Current Opinion in Neurobiology* 13, no. 3 (2003): 372–383.

Hebb, Donald O. *The Organization of Behavior: a Neuropsychological Theory*. New York: Wiley, 1949 (repr. Lawrence Erlbaum Associates, 2002).

Heeger, David J. "Poisson Model of Spike Generation." September 5, 2000. https://www.cns.nyu.edu/~david/handouts/poisson.pdf.

Helmstaedter, Moritz, Kevin L. Briggman, Srinivas C. Turaga, Viren Jain, H. Sebastian Seung, and Winfried Denk. "Connectomic Reconstruction of the Inner Plexiform Layer in the Mouse Retina." *Nature* 500, no. 7461 (2013): 168.

Hendelman, Walter J. *Atlas of Functional Neuroanatomy.* Boca Raton, FL: Taylor and Francis, 2016.

Herculano-Houzel, Suzana. "Scaling of Brain Metabolism with a Fixed Energy Budget per Neuron: Implications for Neuronal Activity, Plasticity and Evolution." *PLoS One* 6, no. 3 (2011).

Hindman, Joseph, Mark D. Bowren, Joel Bruss, Brad Wright, Joel C. Geerling, and Aaron D. Boes. "Thalamic Strokes That Severely Impair Arousal Extend into the Brainstem." *Annals of Neurology* 84, no. 6 (2018): 926–930.

Hobbes, Thomas. *Leviathan.* London: Andrew Crooke, 1651.

Hockney, David. *Secret Knowledge: Rediscovering the Lost Techniques of the Old Masters.* New York: Viking, 2001.

Hoffman, Donald. The *Case Against Reality: How Evolution Hid the Truth from Our Eyes.* New York: Norton, 2011.

Hofman, Michel A. "Energy Metabolism, Brain Size and Longevity in Mammals." *Quarterly Review of Biology* 58, no. 4 (1983): 495–512.

Horton, Jonathan C., and Daniel L. Adams. "The Cortical Column: A Structure Without a Function." *Philosophical Transactions of the Royal Society B: Biological Sciences* 360, no. 1456 (2005): 837–862.

Horvát, Szabolcs, Răzvan Gămănuț, Mária Ercsey-Ravasz, Loïc Magrou, Bianca Gămănuț, David C. Van Essen, Andreas Burkhalter, Kenneth Knoblauch, Zoltán Toroczkai, and Henry Kennedy. "Spatial Embedding and Wiring Cost Constrain the Functional Layout of the Cortical Network of Rodents and Primates." *PLoS Biology* 14, no. 7 (2016): e1002512.

Horvath, Steve, and Kenneth Raj. "DNA Methylation-Based Biomarkers and the Epigenetic Clock Theory of Ageing." *Nature Reviews Genetics* 19, no. 6 (2018): 371.

Hromadka, T., M. R. Deweese, and A. M. Zador. "Sparse Representation of Sounds in the Unanesthetized Auditory Cortex." *PLoS Biology* 6, no. 1 (2008): e16.

Hubel, David H., and Torsten N. Wiesel. "Receptive Fields of Single Neurones in the Cat's Striate Cortex." *Journal of Physiology* 148, no. 3 (1959): 574–591.

Hubel, David H., and Torsten N. Wiesel. "Receptive Fields, Binocular Interaction and Functional Architecture in the Cat's Visual Cortex." *Journal of Physiology* 160, no. 1 (1962): 106–154.

Humphries, Mark D., Kevin Gurney, and Tony J. Prescott. "The Brainstem Reticular Formation Is a Small-World, Not Scale-Free, Network." *Proceedings of the Royal Society B: Biological Sciences* 273, no. 1585 (2006): 503–511.

Hunnicutt, Barbara J., Bart C. Jongbloets, William T. Birdsong, Katrina J. Gertz, Haining Zhong, and Tianyi Mao. "A Comprehensive Excitatory Input Map of the Striatum Reveals Novel Functional Organization." *Elife* 5 (2016): e19103.

Hütt, Marc-Thorsten, Mitul K. Jain, Claus C. Hilgetag, and Annick Lesne. "Stochastic Resonance in Discrete Excitable Dynamics on Graphs." *Chaos, Solitons and Fractals* 45, no. 5 (2012): 611–618.

Hyvärinen, Aapo, Jarmo Hurri, and Patrick O. Hoyer. *Natural Image Statistics: A Probabilistic Approach to Early Computational Vision.* Berlin: Springer Science and Business Media, 2009.

Innocenti, Giorgio M., Marie Carlén, and Tim Dyrby. "The Diameters of Cortical Axons and Their Relevance to Neural Computing in Axons and Brain Architecture." In *Axons and Brain Architecture*, ed. Kathleen Rockland, 317–355. New York: Academic Press, 2016.

Jiang, Jiefeng, Jeffrey Beck, Katherine Heller, and Tobias Egner. "An Insula-Frontostriatal Network Mediates Flexible Cognitive Control by Adaptively Predicting Changing Control Demands." *Nature Communications* 6 (2015): 8165.

Jonas, Eric, and Konrad Paul Kording. "Could a Neuroscientist Understand a Microprocessor?" *PLoS Computational Biology* 13, no. 1 (January 2017): e1005268.

Kaas, Jon H. "Reconstructing the Organization of Neocortex of the First Mammals and Subsequent Modifications." In *Evolution of Nervous Systems*, ed. Jon H. Kaas and Leah A. Krubitzer. Vol. 3. Amsterdam: Academic Press, 2007.

Kale, Penelope, Andrew Zalesky, and Leonardo L. Gollo. "Estimating the Impact of Structural Directionality: How Reliable Are Undirected Connectomes?" *Network Neuroscience* 2, no. 2 (2018): 259–284.

Katz, Barrett, Ronald B. Melles, Michael R. Swenson, and Jerry A. Schneider. "Photic Sneeze Reflex in Nephropathic Cystinosis," *British Journal of Ophthalmology* 74, no. 12 (1990): 706–708.

Kavanau, J. Lee. "Conservative Behavioural Evolution, the Neural Substrate." *Animal Behaviour* 39, no. 4 (1990): 758–767.

Kay, Kendrick N., Thomas Naselaris, Ryan J. Prenger, and Jack L. Gallant. "Identifying Natural Images from Human Brain Activity." *Nature* 452, no. 7185 (2008): 352.

Klamt, Steffen, Utz-Uwe Haus, and Fabian Theis. "Hypergraphs and Cellular Networks." *PLoS Computational Biology* 5, no. 5 (2009).

Keller, Georg B., Tobias Bonhoeffer, and Mark Hübener. "Sensorimotor Mismatch Signals in Primary Visual Cortex of the Behaving Mouse." *Neuron* 74, no. 5 (2012): 809–815.

Kerszberg, Michel. "Genes, Neurons and Codes: Remarks on Biological Communication." *Bioessays* 25, no. 7 (2003): 699–708.

Khaligh-Razavi, Seyed-Mahdi, and Nikolaus Kriegeskorte. "Deep Supervised, but Not Unsupervised, Models May Explain IT Cortical Representation." *PLoS Computational Biology* 10, no. 11 (2014).

Khodagholy, Dion, Jennifer N. Gelinas, and György Buzsáki. "Learning-Enhanced Coupling Between Ripple Oscillations in Association Cortices and Hippocampus." *Science* 358, no. 6361 (2017): 369–372.

Kleinfeld, David, Lan Luan, Partha P. Mitra, Jacob T. Robinson, Rahul Sarpeshkar, Kenneth Shepard, Chong Xie, and Timothy D. Harris. "Can One Concurrently Record Electrical Spikes from Every Neuron in a Mammalian Brain?" *Neuron* 103, no. 6 (2019): 1005–1015.

Kleinrock, Leonard. *Queueing Systems, Vol II: Computer Applications.* New York: Wiley, 1976.

Koch, Christof and Tomaso Poggio. "Biophysics of Computation: Neurons, Synapses and Membranes." In *Synaptic Function*, ed. G. M. Edelman, W. E. Gall, and W. M. Cowan (New York: Wiley, 1987), 637–697.

Konorski, Jerzy. *Conditioned Reflexes and Neuron Organization.* Translated from the Polish. under the author's supervision. Cambridge: Cambridge University Press, 1948.

Kopell, Nancy J., Howard J. Gritton, Miles A. Whittington, and Mark A. Kramer. "Beyond the Connectome: The Dynome." *Neuron* 83, no. 6 (2014): 1319–1328.

Kreiman, Gabriel. "Neural Coding: Computational and Biophysical Perspectives." *Physics of Life Reviews* 1, no. 2 (2004): 71–102.

Kuno, Motoy. *The Synapse.* New York: Oxford University Press, 1995.

Kussul, E., T. Baidyk, L. Kasatkina, and V. Lukovich. "Rosenblatt Perceptrons for Handwritten Digit Recognition." In *IJCNN'01, International Joint Conference on Neural Networks Proceedings*, 1516–1520. Vol. 2. New York: Institute of Electrical and Electronics Engineers, 2001.

Lake, Brenden M., Tomer D. Ullman, Joshua B. Tenenbaum, and Samuel J. Gershman. "Building Machines That Learn and Think like People." *Behavioral and Brain Sciences* 40 (2017).

Laughlin, Simon. "A Simple Coding Procedure Enhances a Neuron's Information Capacity." *Zeitschrift Für Naturforschung C* 36, nos. 9–10 (1981): 910–912.

Learned-Miller, Erik, Gary B. Huang, Aruni Roy Chowdhury, Haoxiang Li, and Gang Hua. "Labeled Faces in the Wild: A Survey." In *Advances in Face Detection and Facial Image Analysis*, 189–248. Cham, Switzerland: Springer, 2016.

Leibniz, G. W. *The Principles of Philosophy, Or the Monadology*. In *Philosophical Essays*, ed. and trans. R. Ariew and D. Garber, 213–225. London: Hackett, 1989. (Original Work Published 1714)

Leiner, Barry M., Vinton G. Cerf, David D. Clark, Robert E. Kahn, Leonard Kleinrock, Daniel C. Lynch, Jon Postel, Larry G. Roberts, and Stephen Wolff. *A Brief History of the Internet*. Internet Society, 1997. https://www.internetsociety.org/internet/history-internet/brief-history-internet/.

Lennie, Peter. "The Cost of Cortical Computation." *Current Biology* 13, no. 6 (2003): 493–497.

Levitt, Jonathan B., and Jennifer S. Lund. "Intrinsic Connections in Mammalian Cerebral Cortex." In *Cortical Areas: Unity and Diversity*, ed. A. Schüz and R. Miller, 145–166. Boca Raton, FL.: CRC Press, 2002.

Levy, William B., and Robert A. Baxter. "Energy Efficient Neural Codes." *Neural Computation* 8 (1996): 531–543.

Lewicki, Michael S. "Efficient Coding of Natural Sounds." *Nature Neuroscience* 5, no. 4 (2002): 356.

Lieberman, Philip. *The Unpredictable Species: What Makes Humans Unique*. New York: Princeton University Press, 2013.

Llano, Daniel A., and S. Murray Sherman. "Differences in Intrinsic Properties and Local Network Connectivity of Identified Layer 5 and Layer 6 Adult Mouse Auditory Corticothalamic Neurons Support a Dual Corticothalamic Projection Hypothesis." *Cerebral Cortex* 19, no. 12 (2009): 2810–2826.

Lowney, Charles. "Rethinking the Machine Metaphor Since Descartes: On the Irreducibility of Bodies, Minds, and Meanings." *Bulletin of Science, Technology and Society* 31, no. 3 (2011): 179–192.

Lythgoe, John Nicholas. *Ecology of Vision*. London: Clarendon Press, 1979.

Mahadevan, Arun S., Nicolas E. Grandel, Jacob T. Robinson, and Amina A. Qutub. "Living Neural Networks: Dynamic Network Analysis of Developing Neural Progenitor Cells." *bioRxiv* (2017): 055533.

Maier-Hein, Klaus H., Peter F. Neher, Jean-Christophe Houde, Marc-Alexandre Côté, Eleftherios Garyfallidis, Jidan Zhong, Maxime Chamberland, et al. "The Challenge of Mapping the Human Connectome Based on Diffusion Tractography." *Nature Communications* 8, no. 1 (2017): 1349.

Marcus, Gary. "Deep Learning: A Critical Appraisal." ArXiv: 1801 00631 (2018).

Marcus, Gary. "Face It, Your Brain Is a Computer." *New York Times*, June 28, 2015. https://www.nytimes.com/2015/06/28/opinion/sunday/face-it-your-brain-is-a-computer.html.

Markov, Nikola T., Mária M. Ercsey-Ravasz, A. R. Ribeiro Gomes, Camille Lamy, Loic Magrou, Julien Vezoli, P. Misery, et al. "A Weighted and Directed Interareal Connectivity Matrix for Macaque Cerebral Cortex." *Cerebral Cortex* 24, no. 1 (2014): 17–36.

Markov, Nikola T., P. Misery, A. Falchier, C. Lamy, J. Vezoli, R. Quilodran, M. A. Gariel, et al. "Weight Consistency Specifies Regularities of Macaque Cortical Networks." *Cerebral Cortex* 21, no. 6 (2011): 1254–1272.

Marr, David. *Vision: A Computational Investigation into the Human Representation and Processing of Visual Information*. Cambridge, MA: MIT Press, 1982.

Marshel, James H., Yoon Seok Kim, Timothy A. Machado, Sean Quirin, Brandon Benson, Jonathan Kadmon, Cephra Raja, et al. "Cortical Layer–Specific Critical Dynamics Triggering Perception." *Science* 365, no. 6453 (2019): Eaaw5202.

Matsuda, Seiji, Naoto Kobayashi, Takehiro Terashita, Tetsuya Shimokawa, Kazuhiro Shigemoto, Katsumi Mominoki, Hiroyuki Wakisaka, et al. "Phylogenetic Investigation of Dogiel's Pericellular Nests and Cajal's Initial Glomeruli in the Dorsal Root Ganglion." *Journal of Comparative Neurology* 491, no. 3 (2005): 235.

McCulloch, Warren S. "The Brain as a Computing Machine." *Electrical Engineering* 68 (1949): 492–497.

McCulloch, Warren S., and Walter Pitts. "A Logical Calculus of the Ideas Immanent in Nervous Activity." *The Bulletin of Mathematical Biophysics* 5, no. 4 (1943): 115–133.

McGee, Patrick. "Elon Musk-Backed Neuralink Unveils Brain-Implant Technology." *Financial Times*, July 17, 2019. https://www.ft.com/content/144ba3b4-A85a-11e9-984c-fac8325aaa04.

Meytlis, Marsha, Zachary Nichols, and Sheila Nirenberg. "Determining the Role of Correlated Firing in Large Populations of Neurons Using White Noise and Natural Scene Stimuli." *Vision Research* 70 (2012): 44–53.

Milo, Daniel S. *Good Enough.* Cambridge, MA: Harvard University Press, 2019.

Ming, Vivienne. "Human Insight Remains Essential to Beat the Bias of Algorithms." *Financial Times*, December 3, 2019. https://www.ft.com /content/59520726-D0c5-11e9-B018-Ca4456540ea6.

Mišić, Bratislav, Olaf Sporns, and Anthony R. McIntosh. "Communication Efficiency and Congestion of Signal Traffic in Large-Scale Brain Networks." *PLoS Computational Biology* 10, no. 1 (2014).

Mišić, Vojislav B., and Jelena Mišić. *Machine-to-Machine Communications: Architectures, Technology, Standards, and Applications.* Boca Raton, FL: CRC Press, 2014.

Młynarski, Wiktor, and Josh H. McDermott. "Ecological Origins of Perceptual Grouping Principles in the Auditory System." *Proceedings of the National Academy of Sciences* 116, no. 50 (2019): 25355–25364.

Mohan, Vishwanathan, Ajaz Bhat, and Pietro Morasso. "Muscleless Motor Synergies and Actions Without Movements: From Motor Neuroscience to Cognitive Robotics." *Physics of Life Reviews* 30 (2018).

Möller, Caroline, Jörg Lücke, Junmei Zhu, Pedro M. Faustmann, and Christoph Von Der Malsburg. "Glial Cells for Information Routing?" *Cognitive Systems Research* 8, no. 1 (2007): 28–35.

Morquette, Philippe, Dorly Verdier, Aklesso Kadala, James Féthière, Antony G. Philippe, Richard Robitaille, and Arlette Kolta. "An Astrocyte-Dependent Mechanism for Neuronal Rhythmogenesis." *Nature Neuroscience* 18, no. 6 (2015): 844.

Murgia, Madhumita. "Why Some AI Research May Be Too Dangerous to Share." *Financial Times* June 19, 2019. https://www.ft.com/content/131f0430 -9159-11e9-B7ea-60e35ef678d2.

Myers, Frederic W. H. "Multiplex personality." *Proceedings of the Society for Psychical Research* 4 (1886–87), 496–514.

Nádasdy, Zoltán, Hajime Hirase, András Czurkó, Jozsef Csicsvari, and György Buzsáki. "Replay and Time Compression of Recurring Spike Sequences in the Hippocampus." *Journal of Neuroscience* 19, no. 21 (1999): 9497–9507.

Narayanan, Arvind. "How to Recognize AI Snake Oil." Princeton University, Center for Information Technology Policy. https://www.cs.princeton.edu/~arvindn/talks/MIT-STS-AI-snakeoil.pdf.

Nelson, Mark E., and James M. Bower. "Brain Maps and Parallel Computers." *Trends in Neurosciences* 13, no. 10 (1990): 403–408.

Newbold, Dillan J., Timothy O. Laumann, Catherine R. Hoyt, Jacqueline M. Hampton, David F. Montez, Ryan V. Raut, Mario Ortega et al. "Plasticity and Spontaneous Activity Pulses in Disused Human Brain Circuits." *Neuron* 107, no. 3 (2020): 580–589.

Newman, Eric, et al., eds. *The Beautiful Brain: The Drawings of Santiago Ramón Y Cajal.* New York: Abrams, 2017.

Noori, Hamid R., Judith Schöttler, Maria Ercsey-Ravasz, Alejandro Cosa-Linan, Melinda Varga, Zoltan Toroczkai, and Rainer Spanagel. "A Multiscale Cerebral Neurochemical Connectome of the Rat Brain." *PLoS Biology* 15, no. 7 (2017): e2002612.

O'Connell, Mark. *To Be a Machine: Adventures Among Cyborgs, Utopians, Hackers, and the Futurists Solving the Modest Problem of Death.* New York: Anchor, 2018.

Oh, Seung Wook, Julie A. Harris, Lydia Ng, Brent Winslow, Nicholas Cain, Stefan Mihalas, Quanxin Wang, et al. "A Mesoscale Connectome of the Mouse Brain." *Nature* 508, no. 7495 (2014): 207.

Olshausen, Bruno A., Charles H. Anderson, and David C. Van Essen. "A Neurobiological Model of Visual Attention and Invariant Pattern Recognition Based on Dynamic Routing of Information." *Journal of Neuroscience* 13, no. 11 (1993): 4700–4719.

Olshausen, Bruno A., and David J. Field. "Emergence of Simple-Cell Receptive Field Properties by Learning a Sparse Code for Natural Images." *Nature* 381, no. 6583 (1996): 607–609.

Olshausen, Bruno A., and David J. Field. "What Is the Other 85 Percent of V1 Doing?" In *23 Problems in Systems Neuroscience*, ed. J. Leo Van Hemmen and Terrence J. Sejnowski, 182–211. New York: Oxford University Press, 2005.

O'Neill, Cathy. *Weapons of Math Destruction: How Big Data Increases Inequality and Threatens Democracy.* New York: Broadway Books, 2017.

Ovsepian, Saak V. "The Dark Matter of the Brain." *Brain Structure and Function* 224, no. 3 (2019): 973–983.

Palmigiano, Agostina, Theo Geisel, Fred Wolf, and Demian Battaglia. "Flexible Information Routing by Transient Synchrony." *Nature Neuroscience* 20, no. 7 (2017): 1014.

Pastor-Satorras, Romauldo, and Alessandro Vespignani. *Evolution and Structure of the Internet: A Statistical Physics Approach.* Cambridge: Cambridge University Press, 2009.

Pavlov, Ivan P. *Conditioned Reflexes: An Investigation of the Physiological Activity of the Cerebral Cortex,* trans. G. V. Anrep. Oxford: Oxford University Press, 1927.

Pelkey, James. "Entrepreneurial Capitalism and Innovation: A History of Computer Communications: 1968–1988." http://www.historyofcomputer communications.info/book/2/2.5%20donald%20davies-65-66.html.

Poggio, Tomaso. "Routing Thoughts." Working Paper No. 258, MIT AI Laboratory, Cambridge, MA, 1984.

Poli, Daniele, Srikanth Thiagarajan, Thomas B. Demarse, Bruce C. Wheeler, and Gregory J. Brewer. "Sparse and Specific Coding During Information Transmission Between Co-Cultured Dentate Gyrus and CA3 Hippocampal Networks." *Frontiers in Neural Circuits* 11 (2017): 13.

Priebe, Nicholas J. "Mechanisms of Orientation Selectivity in the Primary Visual Cortex." *Annual Review of Vision Science* 2 (2016): 85–107.

Quammen, David. *The Tangled Tree.* New York: Simon and Schuster, 2018.

Ramón Y Cajal, Santiago. *Advice for a Young Investigator,* trans. Neely Swanson and Larry Swanson. Cambridge, MA: MIT Press, 2004.

Reardon, Sara. "A Giant Neuron Found Wrapped Around Entire Mouse Brain," *Nature* 543, no. 7643 (March 2017): 14–15.

Reichova, Iva, and S. Murray Sherman. "Somatosensory Corticothalamic Projections: Distinguishing Drivers from Modulators." *Journal of Neurophysiology* 92, no. 4 (2004): 2185–2197.

Reynolds, John H., and David J. Heeger. "The Normalization Model of Attention." *Neuron* 61, no. 2 (January 2009): 168–185.

Rosenfeld, Amir, Richard Zemel, and John K. Tsotsos. "The Elephant in the Room." *ArXiv: 1808.03305* (2018).

Sacks, Oliver. "In the River of Consciousness." *New York Review of Books* 51, no. 1 (2004): 41–45.

Sardi, Shira, Roni Vardi, Anton Sheinin, Amir Goldental, and Ido Kanter. "New Types of Experiments Reveal That a Neuron Functions as Multiple Independent Threshold Units." *Scientific Reports* 7, no. 1 (2017): 1–17.

Schlegel, Alexander, Prescott Alexander, and Peter U. Tse. "Information Processing in the Mental Workspace Is Fundamentally Distributed." *Journal of Cognitive Neuroscience* 28, no. 2 (2016): 295–307.

Schneider, Gerald E. *Brain Structure and Its Origins: In Development and in Evolution of Behavior and the Mind.* Cambridge, MA: MIT Press, 2014.

Scott, Alwyn. *Neurophysics.* New York: John Wiley, 1977.

Scott, Alwyn. *Stairway to the Mind: The Controversial New Science of Consciousness.* Berlin: Springer, 1995.

Seguin, Caio, Ye Tian, and Andrew Zalesky. "Network Communication Models Improve the Behavioral and Functional Predictive Utility of the Human Structural Connectome." *Network Neuroscience* 4 no. 4 (2020): 980–1006.

Seguin, Caio, Martijn P. Van Den Heuvel, and Andrew Zalesky. "Navigation of Brain Networks." *Proceedings of the National Academy of Sciences* 115, no. 24 (2018): 6297–6302.

Seung, H. Sebastian, and Raphael Yuste. "Neural Networks." In *Principles of Neural Science*, ed. Eric R. Kandel and Thomas J. Jessel. New York: McGraw-Hill, 2013.

Seung, Sebastian. *Connectome: How the Brain's Wiring Makes Us Who We Are.* New York: Houghton Mifflin Harcourt, 2012.

Shannon, Claude E. "The Bandwagon." *IRE Transactions on Information Theory* 2, no. 1 (1956): 3.

Sheheitli, Hiba and Viktor K. Jirsa. "A Mathematical Model of Ephaptic Interactions in Neuronal Fiber Pathways: Could There Be More than Transmission Along the Tracts?" *Network Neuroscience* 4, no. 3 (2020): 595-610.

Sherrington, Charles S. *Man on His Nature.* New York: Macmillan, 1941.

Sherrington, Charles. *Integrative Action of the Nervous System.* Cambridge: Cambridge University Press, 1947.

Shoham, Shy, Daniel H. O'Connor, and Ronen Segev. "How Silent Is the Brain: Is There a 'Dark Matter' Problem in Neuroscience?" *Journal of Comparative Physiology A* 192, no. 8 (2006): 777–784.

Somogyi, P., Z. F. Kisvarday, K. A. C. Martin, and D. Whitteridge. "Synaptic Connections of Morphologically Identified and Physiologically Characterized Large Basket Cells in the Striate Cortex of Cat." *Neuroscience* 10, no. 2 (1983): 261.

Soplata, Austin E., Michelle M. McCarthy, Jason Sherfey, Shane Lee, Patrick L. Purdon, Emery N. Brown, and Nancy Kopell. "Thalamocortical

Control of Propofol Phase-Amplitude Coupling." *PLoS Computational Biology* 13, no. 12 (2017): e1005879.

Spencer, Herbert. *Principles of Psychology.* New York: D. Appleton, 1896.

Spiegelhalter, David. "Should We Trust Algorithms?" *Harvard Data Science Review* 2, no. 1 (2020).

Spillmann, Lothar, Birgitta Dresp-Langley, and Chia-Huei Tseng. "Beyond the Classical Receptive Field: The Effect of Contextual Stimuli." *Journal of Vision* 15, no. 9 (2015): article 7, 1–23.

Sporns, Olaf. *Discovering the Human Connectome.* Cambridge, MA: MIT Press, 2012.

Sporns, Olaf, Giulio Tononi, and Rolf Kötter. "The Human Connectome: A Structural Description of the Human Brain." *PLoS Computational Biology* 1, no. 4 (2005).

Stephan, Klaas Enno. "The History of CoCoMac." *Neuroimage* 80 (2013): 46–52.

Steriade, Mircea, and Denis Paré. *Gating in Cerebral Networks.* Cambridge: Cambridge University Press, 2007.

Sterling, Peter, and Simon Laughlin. *Principles of Neural Design.* Cambridge, MA: MIT Press, 2015.

Striedter, Georg F. *Principles of Brain Evolution.* Sunderland, MA: Sinauer Associates, 2005.

Strubell, Emma, Ananya Ganesh, and Andrew McCallum. "Energy and Policy Considerations for Deep Learning in NLP." *ArXiv: 1906.02243* (2019).

Suárez, Laura E., Ross D. Markello, Richard F. Betzel, and Bratislav Mišić. "Linking Structure and Function in Macroscale Brain Networks." *Trends in Cognitive Sciences* 24, no. 4 (2020): 302–315.

Tang, Evelyn, and Danielle S. Bassett. "Colloquium: Control of Dynamics in Brain Networks." *Reviews of Modern Physics* 90, no. 3 (2018): 031003.

Tang, Shiming, Reinhard Wolf, Shuping Xu, and Martin Heisenberg. "Visual Pattern Recognition in *Drosophila* Is Invariant for Retinal Position." *Science* 305, no. 5686 (2004): 1020–1022.

Tasic, Bosiljka, Zizhen Yao, Lucas T. Graybuck, Kimberly A. Smith, Thuc Nghi Nguyen, Darren Bertagnolli, Jeff Goldy, et al. "Shared and Distinct Transcriptomic Cell Types Across Neocortical Areas." *Nature* 563, no. 7729 (2018): 72.

Tebbich, Sabine, Kim Sterelny, and Irmgard Teschke. "The Tale of the Finch: Adaptive Radiation and Behavioural Flexibility." *Philosophical Transactions of the Royal Society B: Biological Sciences* 365, no. 1543 (2010): 1099–1109.

Ter Wal, Marije, and Paul H. Tiesinga. "Phase Difference Between Model Cortical Areas Determines Level of Information Transfer." *Frontiers in Computational Neuroscience* 11 (2017): 6.

Tilghman, Paul. "If DARPA Has Its Way, AI Will Rule the Wireless Spectrum." *IEEE Spectrum* 56, no. 6 (2019): 28–33.

Tomer, Raju, Li Ye, Brian Hsueh, and Karl Deisseroth. "Advanced CLARITY for Rapid and High-Resolution Imaging of Intact Tissues." *Nature Protocols* 9, no. 7 (2014): 1682.

Tovée, Martin J. "Neuronal Processing: How Fast Is the Speed of Thought?" *Current Biology* 4, no. 12 (1994): 1125–1127.

Trettenbrein, Patrick C. "The Demise of the Synapse as the Locus of Memory: A Looming Paradigm Shift?" *Frontiers in Systems Neuroscience* 10 (2016): 88.

U.S. Census. "Historical Census of Housing Tables: Plumbing Facilities." October 2011. https://www.census.gov/hhes/www/housing/census/historic/plumbing.html.

Valenstein, Elliott S. *The War of the Soups and the Sparks*. New York: Columbia University Press, 2007.

Van Den Heuvel, Martijn P., and Olaf Sporns. "Rich-Club Organization of the Human Connectome." *Journal of Neuroscience* 31, no. 44 (2011): 15775–15786.

Vanrullen, Rufin, Rudy Guyonneau, and Simon J. Thorpe. "Spike Times Make Sense." *Trends in Neurosciences* 28, no. 1 (2005): 1–4.

Varela, Francisco, Jean-Philippe Lachaux, Eugenio Rodriguez, and Jacques Martinerie. "The Brainweb: Phase Synchronization and Large-Scale Integration." *Nature Reviews Neuroscience* 2, no. 4 (2001): 229–239.

Vázquez-Rodríguez, Bertha, Laura E. Suárez, Ross D. Markello, Golia Shafiei, Casey Paquola, Patric Hagmann, Martijn P. Van Den Heuvel, Boris C. Bernhardt, R. Nathan Spreng, and Bratislav Mišić. "Gradients of Structure–Function Tethering Across Neocortex." *Proceedings of the National Academy of Sciences* 116, no. 42 (2019): 21219–21227.

Vilarroya, Oscar. "'Two' Many Optimalities." *Biology and Philosophy* 17, no. 2 (2002): 251–270.

Vinje, William E., and Jack L. Gallant. "Sparse Coding and Decorrelation in Primary Visual Cortex During Natural Vision." *Science* 287, no. 5456 (2000): 1273–1276.

Vlastis, Anna. "Tech Metaphors Are Holding Back Brain Research." *Wired*, June 22, 2017. https://www.wired.com/story/tech-metaphors-are-holding-back-brain-research/.

Vogels, Tim P. and L. F. Abbott. "Gating Multiple Signals Through Detailed Balance of Excitation and Inhibition in Spiking Networks." *Nature Neuroscience* 12, no. 4 (2009): 483–491.

Von Melchner, Laurie, Sarah L. Pallas, and Mriganka Sur. "Visual Behaviour Mediated by Retinal Projections Directed to the Auditory Pathway." *Nature* 404, no. 6780 (2000): 871–876.

Wang, Samuel S-H. "Functional Tradeoffs in Axonal Scaling: Implications for Brain Function." *Brain, Behavior and Evolution* 72, no. 2 (2008): 163.

Waxman, Stephen G. "Regional Differentiation of the Axon: A Review with Special Reference to the Concept of the Multiplex Neuron." *Brain Research* 47, no. 2 (1972): 269–288.

Wegner, Daniel M. "A Computer Network Model of Human Transactive Memory." *Social Cognition* 13, no. 3 (September 1995): 319–339.

Weiss, Thomas, Wolfgang H. R. Miltner, Ralph Huonker, Reinhard Friedel, Ingo Schmidt, and Edward Taub. "Rapid Functional Plasticity of the Somatosensory Cortex After Finger Amputation." *Experimental Brain Research* 134, no. 2 (2000): 199–203.

Willmore, Benjamin, and David J. Tolhurst. "Characterizing the Sparseness of Neural Codes." *Network: Computation in Neural Systems* 12, no. 3 (2001): 255–270.

Winer, Gerald A., Jane E. Cottrell, Virginia Gregg, Jody S. Fournier, and Lori A. Bica. "Fundamentally Misunderstanding Visual Perception: Adults' Belief in Visual Emissions." *American Psychologist* 57, nos. 6–7 (2002): 417.

Wiskott, Laurenz. "How Does Our Visual System Achieve Shift and Size Invariance?" In *23 Problems in Systems Neuroscience*, ed. J. Leo Van Hemmen and Terrence J. Sejnowski, 322–340. New York: Oxford University Press, 2005.

Wiskott, Laurenz, and Christoph von der Malsburg. "Face Recognition by Dynamic Link Matching." In *Lateral Interactions in Cortex: Structure and Function*, ed. J. Sirosh, R. Miikkulainen, and Y. Choe. Austin, TX: UTCS Neural Networks Research Group, 1996.

Wolfrum, Philip. "Switchyards-Routing Structures in the Brain." In *Information Routing, Correspondence Finding, and Object Recognition in the Brain*, 69–89. Berlin: Springer, 2010.

Wood, Sophia. "How Mexican App Bridgefy Is Connecting Protesters in Hong Kong." *Latam List*, August 23, 2019. https://latamlist.com/how-mexican-app-bridgefy-is-connecting-protesters-in-hong-kong/.

Xu, C. Shan, Michal Januszewski, Zhiyuan Lu, Shin-Ya Takemura, Kenneth Hayworth, Gary Huang, Kazunori Shinomiya, et al. "A Connectome of the Adult *Drosophila* Central Brain." *bioRxiv* (2020). https://www.biorxiv .org/content/10.1101/2020.01.21.911859v1.full.pdf.

Yamins, Daniel L. K., Ha Hong, Charles F. Cadieu, Ethan A. Solomon, Darren Seibert, and James J. Dicarlo. "Performance-Optimized Hierarchical Models Predict Neural Responses in Higher Visual Cortex." *Proceedings of the National Academy of Sciences* 111, no. 23 (2014): 8619–8624.

Yen, Shih-Cheng, Jonathan Baker, and Charles M. Gray. "Heterogeneity in the Responses of Adjacent Neurons to Natural Stimuli in Cat Striate Cortex." *Journal of Neurophysiology* 97, no. 2 (2007): 1326–1341.

Yoshida, Takashi, and Kenichi Ohki. "Natural Images Are Reliably Represented by Sparse and Variable Populations of Neurons in Visual Cortex." *Nature Communications* 11, no. 1 (2020): 1–19.

Zhu, Xinge, Liang Li, Weigang Zhang, Tianrong Rao, Min Xu, Qingming Huang, and Dong Xu. "Dependency Exploitation: A Unified CNN-RNN Approach for Visual Emotion Recognition." In *Proceedings of the Twenty-Sixth International Joint Conference on Artificial Intelligence*, 2017, 3595–3601.

Zullo, Joseph M., Derek Drake, Liviu Aron, Patrick O'Hern, Sameer C. Dhamne, Noah Davidsohn, Chai-An Mao, et al. "Regulation of Lifespan by Neural Excitation and REST." *Nature* 574, no. 7778 (2019): 359–364.

Zylberberg, Joel, and Michael Robert Deweese. "Sparse Coding Models Can Exhibit Decreasing Sparseness While Learning Sparse Codes for Natural Images." *PLoS Computational Biology* 9, no. 8 (2013): e1003182.

INDEX

acks (acknowledgments):
hypothetical implementation
in brain, 15–16, 114, 158–162, 185,
188–189, 199, 216, 251, on internet,
12, 137, 140, 144, 230
action potential. *see* spikes
addressing, 194, 222–223
Advanced Research Projects
Administration (ARPA), 127,
132–133; ARPANET, 132–133,
137–138, 296n10; DARPA, 136
akinetopsia, 255–256
Al-Haytham, Ibn, 281n12, 281n14
Allen Institute, 112, 115, 159, 186–187,
240–241, 250
ALOHANET, 16, 133–134
amacrine cells, 181–182
Application Layer (internet),
140–142
ARPANET. *see* Advanced Research
Projects Administration:
ARPANET
arthropods, 272–273
artificial intelligence (AI), xiv, 21,
41, 43, 85; and efficient coding
theory, 91, 98; and the internet,
119; implications of the internet

metaphor for, 258–261; and tract
tracing, 241
artificial neural networks: 41, 43, 95;
layers of, 45; and efficient coding,
91; weights, 40–43, 45–47, 235,
282n27, 282n29
Ashby, F. Gregory, 163, 300n21
astrocytes, 177, 221
Atari game system, 277n4
attention, visual, 1–3
auditory system, 94
autism, 241
awareness, 262–264, 273
axon: anatomy and physiology
of 18, 40, 57, 69–70, 149, 169,
172, 250, 296n13; axo-axonal
signaling, 185; dynamic
polarization of, 209–210;
ganglion cell, 87, 182; ephaptic
interactions, 294n19;
myelination, 198, 305n72;
projections to claustrum,
186–187; projections to thalamus,
14–15, 293n17; rewiring, 171;
terminal, 70, 76–77, 175, 177,
209, 249; of thalamic reticular
nucleus, 182; tracing, 102,

axon (*continued*)
106–113, 219–220, 240–241, 245; tree, 174–176, 215, 249, 307n14

Baars, Bernard, 300n25
back-off algorithm: on internet 134–135, 167, 295n7; in neuromorphic computing, 253; possible implementation in brain, 155–158, 215, 270
Bacon, Kevin, 8
Baran, Paul, 127–130, 147, 246
Barlow, Horace, 87, 89–90, 288n21, 289n27
basket cells, 178–179
Bassett, Danielle, 36, 187
Beck, Jeff, 164
Bell, Tony, 17, 290n33
binding (neural), 192, 219
Boehm, Sharla, 130, 246
Borg (*Star Trek: The Next Generation*), 262
botnet, 149
brain: energy use, 11, 149, 286n6; evolution, 86, 217; monkey, 107–111; mouse, 66, 104 106–107, 112, 293n10; network wiring, 53–56, 130; number of neurons in human, 66, 70, 286n3; oscillations, 28, 193–194, 197, 220, 272, 305n69
brainstem, 183–184, 195, 237; and consciousness, 266
Brette, Romain, 264, 312n6
Briggs, Farran, 161
buffering, 185, 215–216, 229–233, 245–247, 250–252, 307n19
Buonomano, Dean, viii
bus (computing), 123, 229–231
Buzsáki, György, 211–216, 302n35, 306n7

cache, 139, 229. *see also* buffering
calcium (ion), 177

camera obscura, 32
cell assembly (neural), 173, 180–182
cell phone, 119
central complex (arthropod brain), 273
central processing unit (CPU), 123, 229–230, 265
cerebral cortex. *see* cortex
checksum, 296n10
chess, 164; and deep learning, 54–55, 300n24
Chinese (language), 92–94, 98, 290n29
chips (computer), 176, 229–230, 252
Churchland, Patricia, 37
circuit-switching, 123–125, 136, 197, 199–203, 225–227, 257
Clark, Wesley, 132
claustrum, 185–187, 195, 237
clocks, biological, 28, 184
cloud computing, 166
cocktail party problem, 212, 306n6
CoCoMac (connectome), 244, 250, 309n9
cognition, 29, 41, 62, 97, 163, 168
cognitive control, 164
cognitive radio, 136
Cole, Michael, 187
collision: of electronic signals, 134–135, 215, 230, 232, 253 295n7; of neural signals, 150, 158, 167, 185, 203–204, 224, 227, 246, 248–251, 310n15
color vision: evolution of, 152, 297n6
communication: asynchronous, 11, 126, 131, 137–138, 194; busy signals, 11–12; channel, 122; design principles, 10, 244; in computer architectures, 229–233; directed, 13; errors in, 11–12; multiplex, 122; networks, 121–123, radio, 16, 133, 135–136; synchronous, 10, 124, 138–139, 194–195

complex cells, 80–84, 287n16, 287n17, 288n18

computation: definition of, viii; metaphor, vii-viii, 37–38, 40–41, 61–65, 67, 84–85, 100, 151, 162, 172, 209–210, 217, 228, 235, 259, 268–269, 312n6, 313n12; parallel, 252

computer metaphor. *see* computation: metaphor

connectomics, ix, 8, 101, 176, 219–220, 234, 240

consciousness, 151, 259, 261–268

controller (computer), 232

convolutional neural networks (CNNs), 48, 98–99

core regions (brain), 103, 106–107, 195, 273

coronavirus (SARS CoV-2), 22, 119, 148, 172

corpus callosum, 198, 305n71, 305n72

cortex: 49, 182, 188–191, 313n1; anterior cingulate area, 114, 294n18; auditory, 113, 159; columns in, 235–236, 252; connectivity of, 109–114; frontal, 20, 110, 111, 183, 187–188; layers of, 236–237, 272; number of neurons, 148; parietal, 187–188; somatosensory, 169, 180; visual, 50–52, 78–84, 94–96, 154, 159, 192, 219, 221, 236

covid. *see* coronavirus

creativity, 266–267

Crick, Francis, 185

crossbar switch, 232–233

Cushman, Fiery, 165

cybernetics, 39, 61

dark matter (neural), 97, 160

DARPA. *see* Advanced Research Projects Agency (ARPA): DARPA

Darwin, Charles, 34, 87

Davies, Donald, 127, 132, 295n6

decision making, xi, 2, 18, 166

deep learning (deep artificial neural networks), 151, 234–235, 259–260, 267; and object recognition 45–48, 98; and brain models, 48–52, 234–235; and brain wiring, 52–54; training, 55; sparseness of, 98

default mode network, 189, 191, 243

deflection routing. *see* routing: deflection

Deisseroth, Karl, 258

dendrites, 18, 39–40, 57, 69–70, 75–76, 102, 174

deoxyribonucleic acid (DNA), 28, 234

Descartes, René, 28–32, 36, 57, 165

development. *see* scaling: developmental

drivers (cortex), 189–190, 197, 222

dualism (Cartesian), 29, 165, 166

dynome, 116–117

echo request, 144, 146

Edelman, Gerald, 302n35

efficient coding theory, 91–97

Einstein, Albert, 165

electroencephalography (EEG), 216–218

electron, 26, 280n1; microscopy, 308n3

email, 130, 140

emergence, 18–19

endosymbiosis, 238

English (language), 92–94

Enlightenment, 27

epilepsy, 193

Ethernet, 134, 142, 295n7

Euler, Leonhard, 253–254

evolution (Darwinian), 206; and Galápagos finches, 152

exchange point, 143, 147, 184
extramission, 33, 281n14

Facebook, 278n8
face perception, 49, 52, 99, 217, 260,
    282n29, 297n3, 297n6
Field, David, 92, 95–96, 98–99,
    289n27, 291n39
Finlay, Barbara, 7, 13
firewall, 224
first-in first-out (queueing), 247, 252
fly: brain, 241, 272–273, 308n3; visual
    system, 91–92, 272, 313n2
fMRI. *see* imaging: functional MRI
Free-Energy Principle, 151, 297n3
Fries, Pascal, 192–194
frontal lobe. *see* cortex: frontal
fruit fly. *see* fly
Fukushima, Kunihiko, 99, 282n26
functional reallocation, 169–171
Furber, Steve, 252

ganglion cells. *see* retinal ganglion
    cells
gate (computing), 176, 180–182
Gazzaniga, Michael, 36
genetic engineering, 102, 256, 258
genomics, 234, 238
glia, 177–178, 215
gliotransmitters, 177–178
Golgi, Camilo, 71–72
Gollisch, Tim, 180–182
graph theory, 253–254; and graph-
    based entropy, 285n53
Gray, Charles, 192

handwriting recognition, 284n44
Hao, Yan, 249, 310n15
Hebb, Donald O., 20
Hebbian learning, 19–20, 167, 231,
    280n24
Herculano-Houzel, Suzana, 286n3

hippocampus, 189, 213, 244, 306n7
Hobbes, Thomas, 33–34
Hodgkin, Alan, 75, 77
Hodgkin-Huxley model, 75–77, 84,
    99, 154, 173, 310n15
Hoffman, Donald, viii
hot potato (routing): *see* routing: hot
    potato
hub (network), 101, 103, 113, 188, 244,
    266; routers, 143
Hubel, David, 4, 77–84, 87, 95,
    287n16, 287n17, 288n18
Human Connectome Project, 116,
    219–220
Huxley, Andrew, 75, 77
hyperedge, 254–255
hypergraph, 254–255

imaging, 183, 220; calcium, 71;
    diffusion, 219–220; magnetic
    resonance (MRI), 71, 239;
    functional MRI, 187–188,
    217–220, 242–243, 255 283n33,
    287n15; structural MRI, 106,
    219–220, 241–243, 307n14, 308n5
immune system, xi–xii
information spreading, 248–252
information theory: *see* Shannon,
    Claude: information theory of
integrate-and-fire (model), 173
internet: dial-up, 197; history of, 7,
    19, 120, 127–130, 132–135, 137–138,
    147–148; size of, 148–149, 267–268;
    of Things, 22, 264; and Web 2.0,
    119
interoperability, 16, 18–19, 108, 138,
    140–141, 143, 166, 171, 244, 268,
    273; and AI, 261
invariance, 5–6; 45, 55–56, 81–83,
    292n7, 313n2
IP address, 146–148
IPv6 (protocol), 296n11

keep-alive messages, 144, 170, 191
Kennedy, Henry, 106–112, 240, 250
kidneys, 298n7
Kleinrock, Leonard, 122, 132–133
Koch, Christof, 185–187
Königsberg: bridges of, 253–254
Konorski, Jerzy, 20
Kording, Konrad, 277n4

lamination (cortex). *see* cortex: layers
    of
lateral geniculate nucleus (LGN). *see*
    thalamus: lateral geniculate nucleus
Laughlin, Simon, 91–92
law of dynamic polarization,
    209–210
learning, 19, 20, 166–167, 188, 191, 204;
    language, 55; bird song, 55
Leibniz, Gottfried, 36–37, 56, 284n47
Licklider, J. C. R., 127
Lieberman, Philip, 9, 163
Linear B, 289n28
locus coeruleus, 183, 303n54
Loewi, Otto, 73–75
Lythgoe, John, 288n21

machine learning. *see* artificial
    intelligence
mail. *see* postal systems
Marcus, Gary, viii
Marder, Eve, 207, 239
Margulis, Lynn, 237–238
Markov, Nikola, 250
Marr, David, 4–5, 48
marsupials, 305n71
McCulloch, Warren, 39–40, 43, 70, 99
McCulloch-Pitts model, 40–42, 75, 173
McIntosh, Andrew, 244, 246
Meister, Markus, 180–182
membrane potential, 231
memory, 166–167, 218, 231; chips,
    229–230; trace, 167

mental workspace, 9, 165, 300n25
mesh networks. *see* network: mesh
message switching, 125
metaphor: clock, 27–28, 228;
    internet, ix, xii, 23–25, 64, 148,
    163, 188, 195; loom, 56–57; mill,
    36–37, 56; plumbing, 29–32, 228;
    and scientific theory, 27, 56–58,
    206–207, 228; technological,
    27–29, 36, 56; telegraph network,
    57–60; railway, 57, 59
microtome, 308n3
Milo, Daniel, 297n7
Ming, Vivienne, 235
Minitel, 121
Mišić, Bratislav, 242, 244–246, 251
modulators (cortex), 189–190, 197, 222
monotremes, 305n71
motor plans (neural), 161, 168
MRI. *see* imaging: magnetic resonance
multiplexing. *see* communication:
    multiplex
multiplex neuron, 174–175
multirealizability (motor systems),
    167–168
Musk, Elon, 268, 312n12
myelination. *see* axon: myelination
Myers, Frederic, 56

nacks (negative acknowledgments),
    137, 162, 199
naks. *see* nacks
Narayanan, Arvind, 259
natural scenes, 289n27, 290n33
net neutrality, 296n12
network: mesh, 266; robustness,
    128–129, 136–137, 147–148; scaling,
    148; science, 52, 283n37; shortest
    paths, 8, 248, 252; small world,
    8, 101, 103, 128, 158, 160, 196, 246;
    social media, 267; topology,
    128–130, 198

neural Darwinism, 302n35
neural networks. *see* artificial neural
    networks
neural re-use, 304n62
neuromorphic computing, 252–253
neuron, viii, x–xi, xiii, 2–3, 5, 8–9,
    11, 13, 18–21, 31, 39–41; doctrine,
    72–73, 75, 100; energy use,
    296n13, number in the brain, 65;
    signaling, 67–69
Neuropixels (electrode), 257
neurotransmitters, 69–70, 74–75,
    175–178; caging of, 311n24
Newton, Isaac, 37
noise, 16, 135, 170; anatomical, 159,
    299n15; instrumental, 299n17;
    physiological, 153, 156, 158
nuclear missile command, 128

object recognition, 3, 45. *see also* deep
    learning: and object recognition
Oh, Seung-Wook, 112, 250
Olshausen, Bruno, 95, 98, 292n7
open shortest path first routing
    (OSPF), 146, 248
optic nerve, 88, 182, 294n19
optimality, 269; and AI, 260–261;
    of brain, 151, 159, 172; in
    computation 151, 155, 260;
    evolutionary, 152–153, 271,
    297n4, 298n7; of network
    communication, 170, 247–248
optogenetics, 228, 256
organelles, 238
oscillations. *see* brain: oscillations
overload, 200

packet (data), 7, 17–19, 137–141, 171,
    232; corruption of, 295n10; loss
    of, 134–137, 139, 215; size of, 295n5
packet-switching, 126, 130–133, 138,
    197–203, 225–227, 252, 257

PageRank (algorithm), 166
pain. *see* touch: pain
paramecium, 245
Pastor-Satorras, Romualdo, 119
pathfinding, 19, 143–146
pattern detectors, 5
Pavlov, Ivan, 58–59, 61
peering (network), 144, 147, 187
Perceptron, 41–45, 95
pericellular nests, 178
photoreceptors, 89, 221
Physical Layer (internet), 140–142,
    196
Pitts, Walter, 39–40, 43, 70, 99
plasticity, 9
plumbing, 29, 31–32, 280n5
Poggio, Tomaso, 1, 3, 6–7
Poisson process, 156–157
postal systems, 125–126, 131–132, 137,
    223; Mongol, 121
posted write (computing), 230
primary visual cortex. *see* V1
pseudoglobal knowledge, 223–224

Quammen, David, 34, 238

radio. *see* communication: radio
Ramón y Cajal, Santiago, 13, 31, 72,
    178, 205–210, 237
RAND Corporation, 127
randomness: and computation, 16,
    154; and information theory, 154;
    in routing protocol, 130, 132, 134,
    270
random walk, 245–250
raphe nuclei, 183
receptive field, 82, 87–88, 96
redundancy, 63, 249–250, 270
refractory period, 297n10
representational exchange, 165
reptiles, 272
request for comment (RFC), 120

rerouting, 169–171, 301n31
reticular formation (brainstem), 183–184
reticular nucleus (thalamus). *see* thalamus: reticular nucleus
reticular theory, 72, 117
retina, 78, 87–90, 169, 180–182, 210, 241, 313n2
retinal ganglion cells, 87–91, 104, 181–182, 221, 289n23
rewiring (neural). *see* functional reallocation
rich club, 103–104
Rosenblatt, Frank, 41–45, 99
router (network), 132, 137, 140–141, 143–144, 170, 176, 232
routing: in brain, 130, 188, 191, 243, 255; deflection, 169; definition of, xii; flexible, 3, 162–168; and energy use, 11; hot potato, 132, 169, 246–248, 252; protocol, 120–123, 126, 162, 270; in single neurons, 174–179; tables, 19, 132, 145–146, 170, 191, 248; queueing, 121, 247, 252; by synchrony, 192–195, 197

Sacks, Oliver, 265–266
scaling: and brain evolution, 22, 197–199, 225–228; of computer architectures, 305n73; developmental, 197, 203–204; of networks, 21–22, 197–203, 225–227; of neural activity, 256–257
schizophrenia, 241
Scott, Alwyn, 174–176
search (internet), 166, 301n28
Sejnowski, Terrence, 37, 290n33
Seung, Sebastian, 241, 262, 291
Shannon, Claude, 62; information theory of, 62–64, 90, 249
Sherman, Murray, 189, 222

Sherrington, Charles, 31, 56–59, 61, 72
Shohamy, Daphna, 187
shortest path. *see* network: shortest paths
simple cells, 80–84, 287n16, 287n17, 288n18
Singer, Wolf, 192
sleep, 192, 270; REM, 262, 265
small world. *see* network: small world
somatosensation. *see* touch
sparse coding. *see* sparseness: sparse coding
sparseness: 92, 270, 290n30, 307n17; in communication, 11; of internet, 137; of networks, 307n17; of neural activity, 11, 93, 94, 97, 160, 171, 173, 188, 226–227, 250, 256; sparse coding, 93–96, 98–99; of written languages, 93
sparsity. *see* sparseness
Spencer, Herbert, 57–58, 61, 128
Spiegelhalter, David, 260
spike (neural), 39–40, 68–70, 77–80, 84, 117, 159–160, 175–176, 193, 217, 219, 252; failures, 250–251; sequences of, 211–213
spikelets, 185
spinal cord, 180, 237
SpiNNaker (chip), 253
spontaneous activity (neural), 170–171, 191, 248
Sporns, Olaf, 101, 116, 244, 246
stack (protocol), 17–18, 139–143, 196, 237, 272
Stanford Research Institute (SRI), 132–133
streaming, 139
striate cortex. *see* V1
striatum, 188
submergence, 18–19

supply chains, 171–172
Sur, Mriganka, 171
surge capacity, 171–172
switchboard, 8, 57, 58–61, 103, 124
synapse, 20, 66–68, 72–75, 175–177, 293n13; tripartite, 177
synchrony (neural), 21, 192–195, 248

task switching, 20–21
Taub, Edward, 169
Taylor, Bob, 127, 132
telcos (telecommunication companies), 143–144
telephony, 124–126, 160, 197, 199, 222–223, 233; party lines in, 278n16
thalamus: anatomy, 114–115; auditory, 171; and consciousness, 266; connectivity, 112–115, 184–185, 187, 195, 236–237, 293n17; as core brain region/hub, 13, 112; lateral dorsal nucleus, 114; lateral geniculate nucleus, 112; lateral posterior nucleus, 112; loops with cortex 13–16, 112, 160–162, 251; reticular nucleus, 113–114, 162, 184–185, 231; ventral medial nucleus, 114; visual processing, 78, 87–88, 104, 112, 171, 182, 221
time sharing, 126–127
touch, 9, 94, 97, 106–107, 114, 162, 168–169, 178; pain, 180
tracer (neural), 102, 104, 108–110, 183, 240–242, 306n12, 309n9; viral vectors, 102
tract tracing. see tracer
transhumanism, 268–269, 313n13
Transport Layer (internet), 140–141, 196
Tse, Peter, 165

uncertainty principle, 288n19
Usrey, Martin, 161
USSR, 128–129

Van Essen, David, 112, 174, 240, 250, 292n7; diagram of, 104–106, 109–110
vegetative state, 262–263
Versailles, Palace of, 29
Vespigniani, Alessandro, 119
video chat, 139
vigilance, 262–264
visual system, 3–5, 45, 48–49, 52, 77–79, 81, 84, 89–90, 94–95, 99, 104–105, 107, 159, 161, 169, 171, 178, 181, 192, 210, 219, 221, 236, 261, 264–266, 272
von der Malsburg, Christoph, 177, 303n45
V1 (primary visual cortex), 14, 78–81, 84, 87, 109–110, 169, 219, 287n16, 287n17, 288n18, 293n15; retinotopy of, 287n15

Walls, Gordon, 293n17
Waxman, Stephen, 174–176
WebRTC, 137
weights (artificial neural networks). see artificial neural networks: weights
weights (neural connectivity), 15, 108–109, 111, 115
Wiener, Norbert, 39
Wiesel, Torsten, 4, 77–84, 87, 95, 287n16, 287n17, 288n18
Wi-Fi, 135–136, 143

yam. see postal system: Mongol

Zalesky, Andrew, 169–170, 244